San Juan Legacy

Capitol City

San Juan Legacy

Life in the Mining Camps

Duane A. Smith

Photographs by John L. Ninnemann

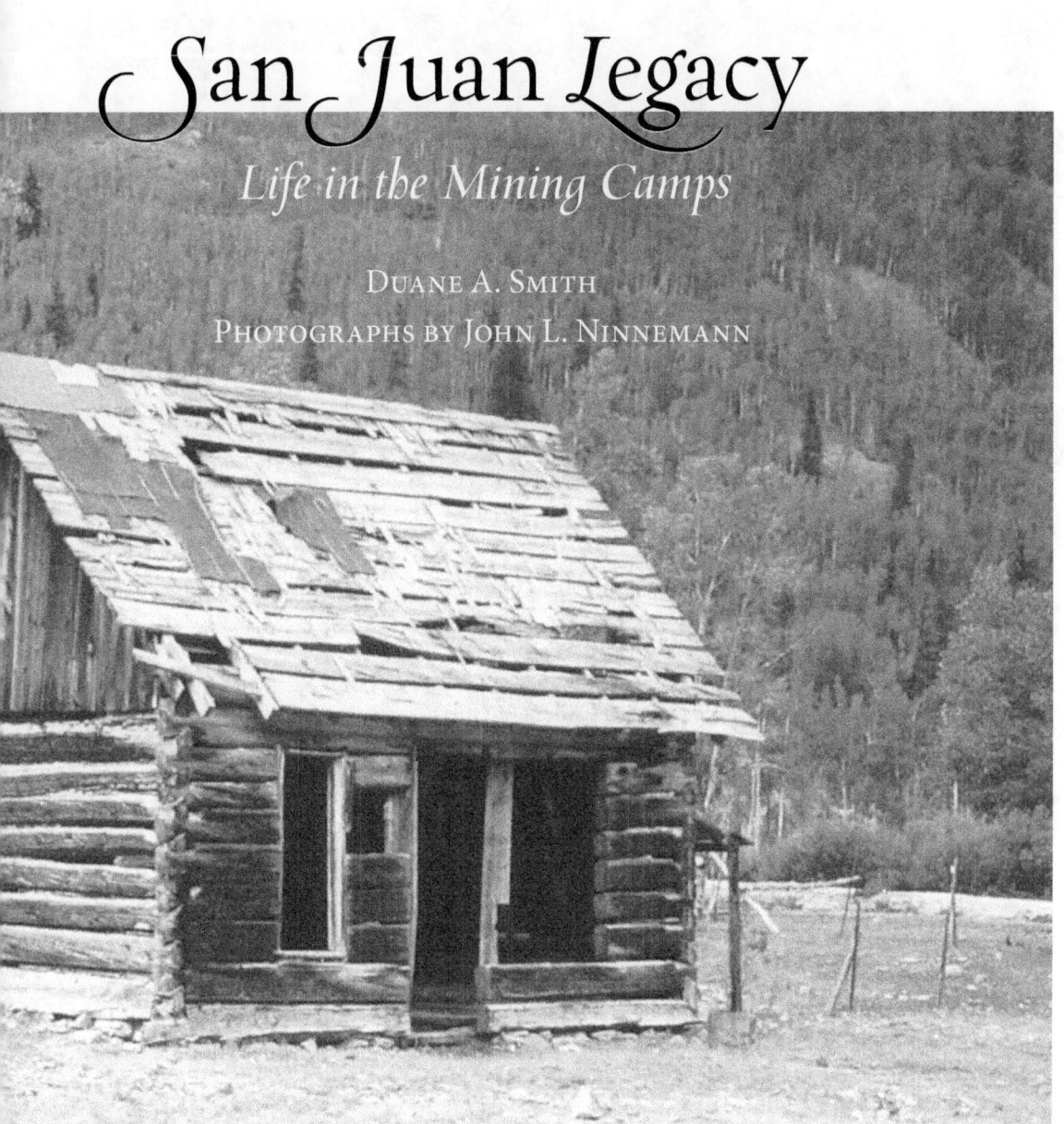

UNIVERSITY OF NEW MEXICO PRESS ～ ALBUQUERQUE

© 2009 by the University of New Mexico Press
All rights reserved. Published 2009
Printed in the United States of America

Library of Congress Cataloging-in-Publication Data
Smith, Duane A.
San Juan legacy : life in the mining camps / Duane A. Smith ;
photographs by John L. Ninnemann.
p. cm.
ISBN 978-0-8263-4650-6 (PBK. : ALK. PAPER)
1. Frontier and pioneer life—San Juan Mountains (Colo. and N.M.)
2. Miners—San Juan Mountains (Colo. and N.M.)—
Social life and customs—19th century.
3. Mining camps—San Juan Mountains (Colo. and N.M.)—History—19th century.
4. Community life—San Juan Mountains (Colo. and N.M.)—History—19th century.
5. Gold mines and mining—San Juan Mountains (Colo. and N.M.)—
History—19th century.
6. Silver mines and mining—San Juan Mountains (Colo. and N.M.)—
History—19th century.
7. San Juan Mountains (Colo. and N.M.) —History, Local.
8. San Juan Mountains (Colo. and N.M.)—Social life and customs—19th century.
9. San Juan Mountains (Colo. and N.M.)—History—19th century.
10. San Juan Mountains (Colo. and N.M.)—Pictorial works.
I. Ninnemann, John L. II. Title.
F782.S18S65 2009
978.8'3031—dc22
2009001513

Book and cover design and type composition by Kathleen Sparkes.
All photographs by John L. Ninnemann. Front cover: the Durango
and Silverton narrow gauge railroad train; back cover: Red Mountain.
The text in this book is Minion OTF PRO, 11/14, 27P.
The display type is Minion and Incognito.
The text ornaments are Caravan.

Dedicated to

Kristi

Scot

Jenifer

Lara

CONTENTS

	Preface ix
PROLOGUE	The Promised Land xi
CHAPTER ONE	Backbone of the Community 1
CHAPTER TWO	No Riding on Sidewalks and License Those Dogs 13
CHAPTER THREE	Raising Hell and a Lot of Other Things 27
CHAPTER FOUR	Transportation Revolution 41
CHAPTER FIVE	"Love Can't Live on Heavy Bread" 57
CHAPTER SIX	Youthful Days, School Days 69
CHAPTER SEVEN	Dentists, Doctors, Disease, and Death 81
CHAPTER EIGHT	"Shall We Gather at the River" or "Shall We Go Straight to Hell" 95
CHAPTER NINE	Age of Joiners 111
CHAPTER TEN	Sin, Sex, and Leisure-Time Pleasures 121
CHAPTER ELEVEN	Culture Arrives in the San Juans 135
CHAPTER TWELVE	"Take Me Out to the Ball Game" 145
EPILOGUE	"Remember me as you pass by" 157
	Index 161

Red Mountain

PREFACE

The English clergyman, novelist, and poet, Charles Kingsley, wrote these words in his poem "Old and New":

So fleet the works of men back to the earth again;
Ancient and holy things fade like a dream.

Nothing could better describe the life and times of the nineteenth-century San Juan mining camps and towns as the first decade of the twenty-first century draws slowly to a close.

It was our desire to re-create for the reader those long-ago times when mining flourished, silver and gold beckoned from the next mountainside or yonder valley, and little communities struggled for existence against long odds. Whether we reached our goal, to recapture as much as possible at this late date the life and times in those camps and towns, we leave to the reader.

We could not have done that without the help and assistance of so many people, museums, and historical societies. We would especially like to thank the staffs of the historical societies and museums in Ouray, Silverton, Telluride, and Durango, Colorado. Without their enthusiastic help and encouragement this project would not have seen the light of day.

We would also like to thank the staff of the University of New Press for their encouragement and professionalism. "Three cheers and a tiger" to our wives, as well, for putting up with another book project.

Ouray

PROLOGUE

The Promised Land

The men would measure in cords the gold they hoped
To find, but the women reckoned by calendars
Of double chins and crow's-feet at the corners
Of their eyes.
—Thomas Hornsby Ferril, "Magenta"

"Westward the star of empire," the pioneers enthusiastically believed as they journeyed west to the promised land, the land of milk and honey. Out they went to a land of wealth in gold and silver, a place to start anew, a wonderful land where the government was willing to give anyone who lived there for five years 160 acres virtually free of charge. Pioneers could start a town, build a railroad, open a store—it was a land of unlimited opportunity for those who went. As well-known newspaperman Horace Greeley advised, "Turn your face to the great West, and there build up a home and fortune."

Miners also joined in the great giveaway. They could claim the land for free, but they had to work it every year or patent and buy their claim for five dollars an acre after surveying it and making improvements.

Nothing motivated these pioneers more than dreams of gold and silver wealth, of "getting rich without working," as the old mining saying went. They rushed in '49 to the California mother-lode country and ten years later to Pike's Peak country. Then, in the 1870s, miners and permanent settlement reached the Colorado San Juans.

There had been miners in these mountains before. In the eighteenth century the Spaniards had left names, legends, and lost mines. In 1861 the rumor of another El Dorado lured the gullible and footloose into Baker's Park, where

Lake City

one day Silverton would grow. They left almost as fast as they came after they found little placer gold, the "poor-man's diggings" of legend.

Still, the legends and stories would not die, and the hopeful prospectors returned in 1869 and the early 1870s. This time mines opened, permanent settlement grew, and the Silver San Juans entered and enjoyed their greatest mining era. Gold production eventually surpassed silver, which had sparked the 1870s excitement. By the time the United States entered World War I in 1917, the San Juans had emerged as one of the country's celebrated mining districts, even considered world-class in production and wealth.

By then the inevitable, gradual decline had set in, but what a generation it had been. San Juan County had produced more than $60 million in gold, silver, copper, lead, and zinc, and San Miguel County topped that with more than $80 million by the 1920s.

The mining camps and towns, which had built the foundation for settlement and development, followed a similar path. Residents of the major towns—Silverton, Ouray, Telluride, Creede, Lake City, and Rico—had worked to re-create the life they had left behind. Even the more numerous and smaller camps tried, as best they could, to emulate their larger neighbors. They all achieved a measure of success, at least in their own eyes. To the visitors and tourists who were starting to arrive, however, the camps, large or small, represented a part of the West of folklore and legend.

Somewhere in and among these images, life and times wove their way along. This, then, is the story that will unfold in the pages that follow.

The book is an attempt to turn back the clock to the nineteenth-century Victorian era (named after Victoria, the long-reigning British queen) and the Edwardian years that followed, as the world rushed toward war and the San Juans slipped into longing for a past that would not come again.

Distance in time may lend a romantic glow, but it does an injustice to the people and their times. They did not fail; they left behind a heritage, a saga of many yesterdays ago. The times, towns, and tribulations have faded into memory, but theirs is a story that cries to be told.

Colorado poet Thomas Hornsby Ferril captured the haunting scene in his poem, "Ghost Town."

> *And here they dug the gold and went away,*
> *Here are the empty houses, hollow mountains,*
> *Even the rats, the beetles and the cattle*
> *That used these houses after they were gone*
> *Are gone: the gold is gone,*
> *There's nothing here,*
> *Only the deep mines crying to be filled.*

Creede

Silverton

San Juan Mining Region 1860–1914

The ubiquitous rolltop desk appeared in offices, homes, and businesses throughout the San Juans. Cluttered or neat, it held a wide variety of papers and other items.

CHAPTER ONE

Backbone of the Community

*Perhaps no town in Colorado can boast of a more reliable
and enterprising class of businessmen, merchants
and bankers [than Silverton].*
—Denver Tribune, September 21, 1881

*One of the great needs of Bachelor at present is a bank.
Merchants require such an institution.*
—Teller Topics, July 22, 1892

Businessmen and businesswomen represented the heart and soul of any mining community. They invested a great deal in developing the town because doing so was good for their businesses, and they hoped for an extended, promising future. These folks could be found involved in almost all aspects of the camp or town's life. The transitory life of the mining West generally did not appeal to them.

They arrived soon after, and sometimes along with, the first prospectors and miners. They or their later counterparts stayed there at the end. Like the miners, they hoped to make a living or, if lucky, their fortunes. In truth, they probably had a better chance of doing so than the miners did.

These businessmen and businesswomen had to know the territory. Prosperity in the mining district created a variety of business successes. Mining towns could support specialized businesses such as newsstands, grocery and clothing stores, butcher shops, and the like. Mining camps supported a general store or two, which carried a variety of goods but offered little selection. The odds of failure, however, remained high.

Businessmen became the mainstays of camp and town government. As discussed in the next chapter, they had the most to gain if a community became permanent and a lot to lose if it failed—more so than most of their customers, who had less invested in the future. They bet on the future.

As shown in the quotation that begins this chapter, the *Denver Tribune* highly praised Silverton's merchants and captured the essence of what those throughout the region strove to achieve. They were, the writer believed, "reliable and enterprising and public spirited," and their dealings "are characterized with fairness and honesty and promptness." These "merchants, bankers and others enjoy a first class reputation at home and abroad."

Main Street businesses got the attention, but quite often the sawmill was one of the early businesses because it answered the immediate needs of the camp and mines. Once a community passed the log-cabin stage, the frame building became a must, or, even better, buildings of brick or stone, which indicated a promising future. A mining community newspaper was remiss if it did not praise such a development.

While not technically part of the business district, another local necessity, boosters claimed, was a mill or smelter. For mining success, an economical method of reducing ore needed to be readily available. A community that gained both a sawmill and ore reduction works held an advantage over competitors. Miners who brought their ore to town or came to purchase lumber were likely to trade there as well.

Each town had its own tributary camps, with no trespassing understood. Yet Ouray and Silverton would sometimes scrap over those camps like cats and dogs. Silverton considered everything north to Animas Forks and Red Mountain within its business domain, while Ouray challenged Silverton coming from the south.

Jealousy between local towns and camps was endemic throughout the nineteenth century. The only time the different factions rallied together came when some outside paper or individual criticized the San Juans. For example, the *Ouray Times* (September 15, 1877) lashed out at Lake City. "Talk about business being dull! Just go over to Lake City and see the long drawn-out countenances of the business men, hear their complaints, notice all the works shut down, and then quit growling about dull times here." Lake City would retaliate with equal vigor.

If a mining district seemed promising, merchants and professionals came to tap the trade that developed. If the district entered a boom period, the growth would seem amazing to visitors. For example, the young Lake City in July 1875 had two restaurants, two groceries, two miners' supply stores, and a couple of saloons. A year later that same business district included fourteen general merchandise and clothing stores, three restaurants, three hotels, two

If one wanted to learn the latest news, the barbershop offered a wonderful opportunity. Telluride went from one in 1884 to three in 1886.

drugstores, two barbershops, nine saloons, two banks, a jewelry store, and a brewery.

Lake City did not prosper as long as many had hoped it would, however, as reflected in an 1879 business census. Two hotels, one bank, one stationary store, two hardware stores, one general merchant, one jeweler, three boot and shoe stores, three meat markets, and only one saloon comprised the business district.

As the years went by, the larger mining communities developed impressive business districts. Telluride in 1899 had two each of banks, hotels, meat markets, and hardware and shoe stores. Six grocery stores, six restaurants, one furniture store, and twenty-two saloons rounded out the district. Silverton and Ouray kept pace except in the number of saloons, and only Durango offered a larger business community.

A farming town would have shown only a fraction of this variety, and it would have developed slowly. In contrast, it would not have seen the ups and downs mining communities experienced.

A little mining camp such as Animas Forks also would have had less variety of businesses than the larger towns. Just getting started in 1878, the camp had a general merchandise store, a miners' supply store, a hotel, one restaurant, and two meat markets. By 1886, though, only one restaurant, a livery stable, and a general store remained.

The omnipresent general store offered few choices but carried a wide range of goods. Stein Brothers' Animas Forks store, for instance, had groceries, dry goods, clothing, crockery, hats, miners' supplies, and "gents goods."

Two other camps displayed the same pattern. In the 1870s, Parrott City's district featured one grocer, a hotel, a shoemaker, and a clothing store, while Capitol City had two general merchants, one blacksmith, and a sawmill. By the early 1880s, Capitol City had three saloons, while

In mine and camp one went to the "mighty smithy" for everything, from horseshoes to the repair of a variety of tools and household items.

the rest of the business district consisted of a general store and a hotel. Parrott City, a pale shadow of its former self, had nearly reached ghost-town status thanks to neighbor and rival Durango. Smaller mining camps' business districts ebbed and flowed yearly and sometimes even seasonally.

A good-sized business district helped boost the community's image and attracted people, investors, and more businesses. For example, when the Bank of Ouray opened in late 1877, it received press coverage in both Ouray and Denver, which must have pleased locals.

Every town had a bank, but few camps did. Durango emerged as the banking and financial hub of the San Juans, at least for the middle and southern portions, and to a lesser degree, for the western side of the region. The records of its oldest bank, First National, present insights into the types of customers served.

If customers were illiterate, several means were used to identify them. A physical description—"colored very black about 5 ft 4," "lost part of forefinger of right hand"—offered the easiest way a customer "can be identified" by a bank employee. Interesting policies included an instruction that "when money is drawn the lady to accompany the gentleman."

A prominent Colorado banking family, the Thatcher brothers, owned the Miners and Merchants Bank in Ouray. Among the services offered were "general banking, discount and collection business." The bank also purchased or would "make advances on gold dust, gold, silver, or base bullion, mattes, concentrates," and so forth. The bank also had an "exchange on all the principal cities of the world."

Unlike the situation in some earlier mining communities of Colorado, there was neither a shortage of currency, nor was gold dust used for exchange purposes. This generation preferred hard currency, so silver dollars and gold coins were very popular. A few businesses did use tokens, which were good for so much in trade or drink or what have you.

In these early years, transportation remained a problem, and blessed was the community that offered easy or all-weather access to the outside world. Goods came in on wagons or pack animals before the railroad arrived, and in the snowy, cold wintertime, passes could be blocked for weeks or months on end. Unless merchants carefully planned ahead, their shelves might become quite bare while prices mounted. Complaints mounted about shortages, but little could be done until conditions improved.

Crusty Mark Twain called the telephone "a time-saving, profanity-breeding, useful invention." San Juaners certainly agreed with his first and third points.

By the century's turn, some merchants were opening stores in more than one town. San Juan Hardware, for example, opened stores in Durango, Silverton, Ouray, and Telluride. Further, it guaranteed its goods at prices "as low as legitimate business affords."

The red-light district, which will be discussed in chapter 10, was a main part of the business world in the camps and towns. It furnished a signpost to a community's prosperity: the more saloons, the better the economy. For more discreet and refined folks, Silverton offered a brewery garden with a lunch counter "ladened with all the delicacies of the season." There

Because of their "pleasant voices," women became centrals. It opened a new field of opportunity for them.

would be "no disturbance on the grounds, officers would be in attendance to keep it quiet."

Before pasteurization became common, allowing beer to be shipped, breweries were common features in communities as well. In 1881, for instance, Silverton and neighboring Howardsville, Durango, and other communities in the San Juans all had them.

The motion-picture theater emerged as part of the new entertainment world in up-to-date, turn-of-the-century communities. The silent one- and two-reel films were wonders of the day. Almost as important was the fact

Some customers were convinced that merchants used crooked scales; most likely they did not. High prices upset more people.

that they signaled that a town was as modern as its counterparts to the east and west.

Another sign of a progressive community soon appeared: the telephone. By the twentieth century, a town without phone connections languished, sadly from the public's point of view, behind the times. Women found work in the telephone business as centrals. It was thought that women's voices were more pleasant than men's, and their hands were smaller so they could better operate switchboards. Electric lights were also a sign of progress.

New might be fashionable, but old standards continued. Rico's *Dolores News* (February 18, 1882) put a significant part of the business community in perspective: "Good hotels and eating houses are a very handsome ornament in mining camps, about the first thing looked for by a newcomer." Lake City understood this in 1901 when it found itself without a good hotel. "Business men ought to get together to get a good hotel man to take hold," urged the *Times*. Why? "It is of vital interest to the well being of Lake City, and means should be used to prevent the slandering and degrading of the community simply because it is way below par in this one respect."

Hotels were indeed a feature of any business district, the fancier the better. They were the places where visitors and investors enjoyed a delicious meal and a comfortable night's sleep. None in the San Juans surpassed Ouray's Beaumont when it opened in 1887. Proud locals claimed it compared "favorably" with the best in Denver. Its "main dining room is the finest in the state without exception."

Its principal rivals were Durango's Strater and Telluride's New Sheridan, both established later. When the latter opened in 1898, the proud *Examiner* gushed, "It is complete in every particular and a credit to the city." Opening-night visitors were treated to "one of the most elegant spreads ever laid in the city."

Mining engineer Eben Olcott, however, stayed in a Silverton hotel in 1880 that sat at the opposite end of the lodging spectrum. "They think you are stuck up if you ask for a bed alone, not to say a room. They open their eyes at your wanting any better place to wash than the common sink of their office."

Restaurants and boardinghouses, in this masculine world, proved necessary. None, however, were larger than "Tortoni of Creede," which had "facilities for feeding 1,000 people daily." With Creede booming, the need certainly existed, but no other community had that size restaurant. Most restaurants were not fancy, but when single men came into town, they offered a better meal than the men probably cooked for themselves.

Women often played a major business role in boardinghouses and rooming houses as well as in other stores in the business district. Many were clerks or waitresses, and some performed other jobs, but few owned businesses.

By the 1890s, mail-order catalogs and people traveling to Denver by train to shop were cutting into local business. Merchants complained, as will be seen in chapter 4, but they could not compete with the variety of goods and the prices offered elsewhere.

Drummers also cut into local business, although some thought they were the harbingers of spring. Traveling salesmen were the brunt of many jokes, but they were also excellent weather vanes of a community's prosperity. The more who arrived, the better the community's prospects.

Local merchants and customers sometimes became their own worst enemies. In 1899, those in Rico were fighting like Kilkenney cats. People and merchants were boycotting each other, according to the local paper, as

Ledgers and a host of documents underlaid the business world of the mining camps. When the alpha and omega of the day was credit, they were essential components.

well as buying goods outside the community that they could buy "at home." The *News-Sun* pleaded for unity, asking folks to "bury the hatchet and all work together," before the situation created "our utter ruin."

Local folk could also undermine the business community. The *Creede Candle* (July 15, 1911) took its readers to task: "For God's sake quit chewing the rag around the streets like an old fish-wife." The next year the *Silverton Standard* hit at "untrue people." The paper admonished, "The prosperity of a town depends chiefly upon the confidence the people have in it."

For both these communities, and their neighbors as well, the boom days lay behind them. Business was not as good as it had been, and prospects were less bright. It seems obvious their residents had become a bit testy and more money conscious.

Prosperity had been the key business word since the 1870s. Small-scale booms and busts had been recurring themes on the local level. This caught many businessmen off guard, but the cycle was part of life in the mining West.

When the 1893 crash occurred, for example, banks throughout the San Juans were hard hit. In those days before bank insurance, depositors stood at the mercy of fate. Every business in the region was affected as a harsh depression settled over the state and region. When the price of silver collapsed, silver mining declined. The ripple effect became widespread. Some camps never recovered.

The high-rolling old days had become history by the 1914 outbreak of the Great War. What the future held, none could tell. Yet the newspapers loved to applaud "three cheers and a tiger" for the businessmen and businesswomen of the San Juans who had persevered over the previous generation and even those who had tried but failed.

Rico's stately courthouse reflected ultimate respectability and lent an air of permanence and stability to the town.

CHAPTER TWO

No Riding on Sidewalks and License Those Dogs

No person shall ride any animals or drive any animals attached to vehicles at a speed exceeding 6 miles per hour.
—Bachelor City Ordinance, July 22, 1892

The Board ordered the ordinance concerning dogs to be enforced strictly by the town constable.
—Silverton town minutes, March 3, 1880

The mayor stated the meeting had been called to consider establishing a chain gang for prisoners. After discussion, no action was taken.
—Rico, Board of Trustees minutes, October 1, 1892

Government, Massachusetts senator Daniel Webster argued, was "the people's government, made for the people, made by the people, and answerable to the people." His contemporary, philosopher Ralph Waldo Emerson, added his thoughts: "The less government we have, the better—the fewer laws, and the less confided power."

Combine those two statements, throw in a little western practicality, and mining-town government emerged. In the rush of urbanization into a new district, some form of government had to emerge to bring stability out of chaos and law and order out of individual desires.

For most smaller camps, the less government the better because creating and maintaining a civil government cost money and time, both of which also weighed against the need to make one's fortune the primary goal in life. Politics remained a second or third consideration. Mining

> *While crime was not a big problem, sobering up drunks was. Jail life must be "very satisfying," observed Mark Twain. Those who come out "don't want to go back."*

towns, however, soon evolved a governmental structure that in many ways reflected that of their counterparts elsewhere in the country.

Local businessmen generally dominated government. For example, in 1892 Creede's mayor was a lawyer, and the town treasurer was a banker. The four aldermen owned mines, a hotel, a mercantile agency, and a lumber company. Not all these men were honest. Telluride's treasurer disappeared in 1910 with his books showing a $36,763 deficit.

These businessmen served on city councils, boards, or as trustees (the titles varied); worked behind the scenes; and helped shape ordinances and tax structure. They had more invested in the community than many of their neighbors did, and their success and profit rested on growth, permanency, and civic aura. Investors, visitors, and settlers were drawn to a "civilized" community more than to one with a wide-open, anything-goes mentality.

Rico's *Dolores News* (September 11, 1879) chastised its readers for not forming a town government. The town site had been surveyed and streets and alleys "properly defined," but people had been "very tardy in the organization of town government." The admonition apparently worked because in December an election was held electing a mayor, a recorder, and four trustees (although the only contests were for the latter four seats). The winners "marched to Fran's Place where toasts were delivered" and then down Glasgow Avenue and "gave three loud and startling cheers for each member of the board." City government had come to Rico.

A typical city government consisted of a mayor, a recorder, a treasurer, and trustees, as in Animas Forks in 1882. A marshal and a fire warden—often the same individual—and a few committees rounded out the official governing body. Creede a decade later had a mayor, a treasurer, aldermen, a marshal, and—because it was larger than Animas Forks—a few more city employees, including a city attorney.

Creede's nearby neighbor, Bachelor, also organized a town government during the exciting days of the 1892 silver rush. The first meetings tell much about the problems a new community faced.

In May, after the election, the *Teller Topics* congratulated the board of trustees for "getting down to business in a substantial way" and "doing much effective work." They also praised them for "getting along." Whether they continued such congeniality has been lost to history. To assure attendance, Telluride, meanwhile, fined trustees ten dollars for missing a meeting.

The first ordinances the board passed reflected local concerns and the way the trustees planned to govern. Among other things, they authorized the appointments of a treasurer, an attorney, a marshal, and a recorder—all to hold office for one year—and defined their "powers and duties." Standing committees would also be appointed annually. Concerned prudently about health, as was every mining community, they hired a town physician and appointed a board of health.

Carrying concealed weapons was prohibited, as were "offenses against public peace" and regulating the speed of "driving and riding animals." The trustees defined nuisances and established fines for them and also established a poll tax (all "able bodied male citizens" either paid a three-dollar poll road tax or spent two days laboring on the roads in the town). If a person could not pay the fine, neighboring Creede allowed him or her to work it off at the rate of two dollars per day. Economy remained a byword for city governments. Creede, for example, boarded prisoners at thirty-five cents per day.

Establishing business licenses (a major revenue generator), writing an ordinance about fire prevention, and establishing yearly appropriations (nine thousand dollars) for the "fiscal year beginning July 19" also took up their time. Whether in Bachelor, Creede, or anywhere else, revenue sources needed to be found.

No town or camp overlooked the importance of business licenses. They were based, however, on respectability, not on revenue generated. For example Telluride's fee for a hotel, restaurant, merchants, and boardinghouses was a total of five dollars each. Gaming tables were five dollars each, dance halls paid twenty-five dollars while posting a bond of five hundred dollars, and theatrical companies were assessed five dollars per day. In 1909, saloons paid six hundred dollars per year, prize fights fifty dollars each, and circuses fifty dollars per day in licensing fees. Fines also helped. By fining gambling and prostitution, then allowing them to flourish and consequently collecting fines every month, city treasuries gained a major source of revenue. That little ploy got some folks upset. Ouray citizens had become aggrieved in 1883 "in regard to the social evil now existing promiscuously throughout town." Lower taxes or not, they wanted it discontinued.

Each city employee had his job defined (initially no women worked in municipal government). For example, the town attorney was to draft all ordinances, leases, and government instruments and act as the town's legal adviser. The town marshal had to post a bond of five hundred dollars and then perform "all duties required by ordinances of the town," including serving as master of the dog pound, and devoting "all his time to the performance of his duties." As the representative of law and order, the marshal had to be "above reproach." Telluride specifically stated that if the marshal was "found gambling his office would be declared vacant."

City government was now launched. Its implementation over the years remained a question, and a look at some of the major communities' city minutes shows why. The issues tackled covered a gamut of concerns that were relevant in the nineteenth and early twentieth centuries.

Most matters—bills, committee reports, budget matters, and citizen concerns—were routine, but occasionally something special happened. Silverton's council had the pleasure of accepting $10,000 from Andrew Carnegie to build a library. An ordinance was approved appropriating $1,250 each year for a library fund, a library board was established, and any person who committed "injury to the library building," refused to return a book, or created a disturbance against the "peace in the library" was liable for a fine of $1 to $100.

Concerns about morality led to a series of ordinances revolving around the red-light district, discussed in chapter 10. Towns such as Telluride also had ordinances directed toward shows, play, concerts, and exhibitions that had the "tendency to be against good morals or decency." Indecent exposure; "indecent books, pictures or other things"; or an "indecent act of behavior" were also prohibited in Creede, Rico, and elsewhere.

Dogs proved a perpetual problem; large numbers ran everywhere. Council after council passed a tax on dogs and demanded that their owners purchase a collar and tag. Dogs without them in Lake City would be impounded for twenty-four hours and "killed if not claimed."

Ouray's dogcatcher was paid $1.50 for each "dog, bitch or whelp" he apprehended. Even with such remuneration, the town had a difficult time finding anyone to take the job. Silverton charged only $1 per dog and ordered the ordinance to be strictly enforced. Usually, these rules proved dead letters unless a hydrophobia outbreak occurred. If vigorously enforced, however, doggie ordinances could rile the taxpayers emotionally.

The marshal had many other duties, including in Ouray rounding up "street loafers, gamblers, trespassers, burglars, lewd women, decoying females and vagrants," which covered a multitude of sins. The bothersome chore of Telluride's marshal was "to see that all small boys and children" were kept "from running at large on streets after" nine at night.

As soon as streetlights appeared, interestingly, trouble started. Lake City in 1911 offered a ten-dollar reward for information "leading to the arrest and conviction of anyone interfering with electric lights, wires or globes." They were tempting targets for kids with new BB guns.

The marshal not only had to check on dogs and "armed" kids, but he also had to maintain law and order. Unlike the fictional or Hollywood

Safes, large and small, were found everywhere. It was not just the outlaw who robbed them. The president of Lake City's bank embezzled forty thousand dollars.

marshal, in Rico in 1912 the real-life one dealt with fighting, obscene language, assault, dog violations, undefined offenses against the peace, and nuisances. Silverton's 1910 crime sheet included drunks, fighting, fast driving, concealed weapons, disorderly conduct, and the all-encompassing disorderly conduct.

Gunfights were rare and robberies only slightly more so. During a camp's or town's birth, lawlessness might briefly get out of hand; all the larger communities of the San Juans went through such periods. Once settlement took hold, however, businessmen became dominant, and people worried about the community's image. Then the drive to create a "civilized" community pushed out the rowdy, law-breaking crowd or limited it to a specific section of town. Carrying of concealed weapons (an ordinance fine) was also rare. Mining camp residents felt little need to carry guns; few even had one except for hunting.

Danger did exist, though, for those who pinned on the badge. Durango and Silverton, for instance, both lost lawmen in the line of duty. Others were wounded, beaten up, or threatened while trying to make an arrest.

Sometimes, economic hard times led to the marshal's salary being reduced or the decision to let the day or night marshal go and have the other one cover both time slots. This obviously cut down on enforcement of ordinances and opened the door for potential trouble. By the time such a step was taken, though, the community was probably no longer particularly attractive for criminals.

Business licenses, or rather their cost, were like the dogs; they never went away. Sometimes local papers addressed the issue, feeling the revenue raised was "but small," or, as others did, chastising the merchants for not paying.

In 1906 Lake City protested the dumping of tailings from the Golden Fleece Mill into the Lake Fork of the Gunnison River because they carried cyanide. Telluride did the same when tailings were dumped into its water supply. An 1885 ordinance threatened a fifty- to three-hundred-dollar fine for anyone "polluting, corrupting or contaminating" Cornet Creek. Such health concerns, however, ran up against a fundamental issue—curbing the mining that had given the town its birth, its life, and its substance. The ordinance did not work as planned, and in 1889 "dead hogs and other refuse" were being dumped into the water.

Fire, one of the great fears of any mining community, received careful attention from the boards. Telluride created its fire department in 1886; its members were volunteers initially, then paid. Silverton, Ouray, and other towns followed that pattern; camps seldom moved beyond volunteers responding to the cry of fire. Meanwhile in the hopes of preventing a conflagration, a series of fire ordinances addressed stovepipes and chimneys,

Nothing scared the residents more than the dreaded cry of fire. Camps and towns all heard it eventually. It was hoped that the fire laddies were up to the challenge.

storing hay and dynamite, and requiring stoves to have a metal sheet or box of sand under them.

The modern world crashed into the towns in a variety of ways. For example, the councils had to consider assessing a "moving-picture" fee. Telluride approved $125 for a photo of the town to be placed in a book on the history of Colorado. Creede built a float for a Denver parade to "represent the resources of Creede."

As the years went by, and particularly after 1900, nearly all San Juan communities found revenues decreasing and the demand for services remaining steady or increasing. By the time of World War I, towns such as Rico, Lake City, and Creede were struggling—reducing salaries and cutting services, for instance—and many small camps gave up even trying to maintain a government. Even getting a quorum for a meeting was sometimes a problem. Lake City tried five times to get a quorum for a meeting in 1909 before the sixth attempt succeeded.

The public generally watched the city fathers with a careful eye. Occasionally they complimented them, as Telluride's *Journal* editor did in 1882, for handling the "affairs of the town and its interests in a very satisfactory and business [like] manner." He then applauded the council "for doing all in its power to promote Telluride's interest and welfare."

More commonly, however, citizen complaints surfaced. Silverton, the *Miner* thought, needed a waterworks during the paper's 1881 drive to make the "town a most attractive and desirable place" to reside. Rico, meanwhile, debated whether the mayor and trustees should be paid when "pressing necessary improvements" were needed. Taxes posed continual vexing problems particularly when a community started to decline. Aroused residents had their big or little concerns. Congregationalists, for instance, became upset when Creede refused to give them free lots for their church.

When the city council permitted a freeze-out game on the city's printing contract, the losing *Herald* rose up in arms. It informed "all fair minded citizens" of Ouray that "much money would be saved if competition were allowed." The rival *Plaindealer* took it all in stride; such was life in local newspaper rivalry. Telluride residents worried about the lack of enforcement of the sidewalk ordinance, the failure of which might make the "city liable to pay damage suits" for falls. Common complaints appeared, including not enforcing the dog ordinance, "wild riding," problems in the red-light district, and a host of other concerns.

Owners of buildings were not in the least happy when their property was destroyed in 1889 by "direction of the fire chief to allay the late fire" in order to prevent that "fire from spreading." It took Rico four months to resolve the issue before one aggrieved owner accepted $250. Nor were some

In courthouses throughout the San Juans some epic and costly legal battles were fought. Add a church and a school, and respectability had arrived.

residents pleased when "dance halls and houses of that class" appeared in their neighborhood. As a result Columbia trustees received petitions both for and against removing dance halls.

Fast riding and driving on roads or over sidewalks aroused continuous complaints. An ordinance had to be passed when drivers of the newfangled automobiles drove down sidewalks. In 1910, Ouray's council ordered that license plates be attached to the rear of every auto, that no one younger than sixteen be allowed to drive, that a ten-mile-per-hour speed limit be observed, and that owners carry "at all times a gong, bell or horn and two lamps." Those would certainly be crucial if drivers persisted in using the sidewalks.

Disgusting smells also upset locals, particularly manure and outhouses. The marshal usually got the job of seeing that owners cleaned up the mess "at once."

An ordinance might lead to unexpected trouble. In an 1898 push for civic improvement, Telluride passed an ordinance prohibiting animals from running at large. The *Examiner* pointed out that this practice left the alleys in "very bad shape" because those "greatest of all scavengers, the burros, are not given a chance to clean them up."

Burros, as cute and helpful as they could be, also brayed and sometimes nearly overran a community. In 1914, Creede became so upset with "stud burros" that the council declared them a nuisance.

A general concern, too, focused on residents' lack of interest in elections. If there were no pressing issues or controversial candidates, the electorate stayed home, much to the concern of editors and folks interested in having the public involved in city government.

Some partisanship existed in local elections, but generally candidates ran on their own reputations. In state and national races, party allegiance played a much larger role.

Colorado and national politics entered into the San Juan political scene every two and four years. Being an isolated and small populated area, the region did not carry much political weight. That did not deter San Juaners however.

In early San Juan politics the Republicans dominated the scene. Civil War memories were fresh, and if one shot against those "damn confederate Southern Democrats," one voted against them too. By the mid-1880s,

They were sometimes "stuffed," "kidnapped," or found empty, but the ballot box symbolized democracy in action.

allegiances had weakened, and in the 1890s, the Populists and Democrats swept the field with their support for increasing the price of silver. The price had been slipping downward, and the Republican-controlled Congress did nothing to heed their cries for sixteen to one, or the price of sixteen ounces of silver equaling one of gold (twenty dollars).

The peak came in the 1896 vote for Democrat William Jennings Bryan, who captured 83 percent of the vote statewide and an even higher percentage in the San Juans. He repeated his success again in 1900. In 1904 Theodore Roosevelt captured the state, though Bryan won again in 1908, but the Republicans did better each year after the silver issue died in 1896.

San Juaners did not enter city government or politics generally for the financial remuneration. For all the worries and responsibilities involved, town salaries were not particularly remunerative. Rico in 1883 paid the mayor $100 and the town clerk $300 per annum and each trustee $5 per meeting, "not to exceed two meetings per month." That same year Ouray paid its marshal $1,200 (a good salary, however, for the time), the treasurer $250, and the town attorney $250 per year.

Hard times, such as the 1893 silver crash, caused a reduction of salaries and services. Ouray reduced salaries by 20 percent, dispensed with the street sprinkler, and cut off streetlights. The situation was not helped when the First National Bank, where the city "deposited its funds," closed it doors in August.

Times got better, and in 1900 Ouray's treasurer received $300, the clerk $400, and the council members split $1,200 evenly. The marshal, however, received the same salary with increased responsibilities, including serving as street supervisor, superintendent of the waterworks, and fire warden. Silverton in 1902 paid its marshal $1,080, the city attorney $770, and the town clerk $425 plus $1 per liquor license and show, and 50¢ for all other licenses.

In the often fast-changing world of San Juan mining communities, the problems facing town governments also changed quickly, along with the population and local mining fortunes. One has to admire the people who worked to create a government base. Not all of them succeeded, but that many did is admirable. In the end, at least for a brief while, the towns re-created typical governments for the time and laid a basis for the future.

"It is a newspaper's duty to print the news and raise hell."
San Juan newspapers lived up to this adage, quite often in spades.

CHAPTER THREE

Raising Hell and a Lot of Other Things

It is a newspaper's duty to print the news and raise hell.
—Chicago Times (1861)

*All successful newspapers are ceaselessly querulous and bellicose.
They never defend anyone or anything if they can help it; if the job is forced
upon them, they tackle it by denouncing someone or something else.*
—Henry Louis Mencken

*There are three things that no one can do to the entire satisfaction of
anyone else: make love, poke the fire, and run a newspaper.*
—William Allen White

These three quotes pretty much encompass the essence of San Juan newspapers, although Mencken's observation was a bit harsh. These gadflies defended, challenged, and promoted their communities; reported news; created news; stirred up the scene; jabbed political opponents; and, it was hoped, provided a lively paper for readers. They also alternately amused, informed, pleased, chastised, angered, challenged, and provoked their readers while continually reminding them to subscribe and advertise their businesses. For papers to be successful and survive, the editors had to tackle this multitude of tasks with unsurpassed vigor or their papers might not last a season.

Newspapers were one of the first things every mining camp and town longed to have. They were what every community needed if it wanted to exist for long in the world of neighborhood fights, intermountain spats, mining rivalries, and regional jealousies while continuing to promote and advance itself on the larger scene.

The newspaper exemplified a camp's or town's status. Every town worth its salt aspired to have its own spokesman, or several, which indicated booming prosperity. Newspapers bowed to their readers throughout the San Juans, thrived during the booming 1870s and 1880s, and declined in numbers thereafter. Durango had five papers within a year of its 1880 birth. Lake City gained three in the 1870s and five more in the 1880s. In the 1880s papers popped up everywhere—Red Mountain had five, Ouray four, Rico six, Telluride five, and Silverton an almost unbelievable thirteen. Little mining camps, except for Red Mountain, generally claimed only one; they included Ames, Ironton, and Animas Forks, and, in the 1890s, Amethyst and Spar City.

This track record of brief existence and failure does not tell the whole story. Excessive enthusiasm led to fluctuations, as did the lack of growth or mining development, poor editing and dull writing or lack of understanding of the town's or camp's situation, and too much competition for available readers and advertisers. All papers helped their communities, if only briefly.

The editor of a mining-camp newspaper was one of the most important people in town. Samuel Clemens (better known as Mark Twain), onetime editor of Virginia City, Nevada's, *Territorial Enterprise*, explained it this way: "In Nevada for a time, the lawyer, the editor, and banker, the chief desperado, the chief gambler, and the saloonkeeper occupied the same level in society, and it was the highest."

He went on to describe, in his classic story of the mining West, *Roughing It*, the editor's duties. Twain, a young man when he assumed the lofty position, was told to "go all over town and ask all sorts of people all sorts of questions, make notes on the information gained, and write them out for publication." He was not above stretching the truth or fabricating a little, nor were some of his San Juan contemporaries. Editor Twain concluded: "However, as I grew better acquainted with the business and learned the run of the sources of information I ceased to require the aid of fancy to any large extent, and became able to fill my columns without diverging noticeably from the domain of fact."

The key to a newspaper's success focused on the editor, as Twain knew. He had to "rustle" the news, compose each edition with the help of a small staff at best (most of the San Juan papers were initially weeklies), find advertisers and subscribers, and survive in a challenging world.

Ouray and Durango's David Day was a perfect match for Clemens. Civil War veteran and Medal of Honor recipient, Day had founded the *Solid*

Muldoon in 1879. As his wife said, "He attacked anyone, even his friends if he thought they were wrong." Victoria added that her husband did not know "what fear was . . . he wasn't afraid of the Devil."

Day stirred the pot as few editors dared. Durango Democrats brought him to town in 1892 to defend and promote their party against all the local Republican papers. They had caught the tiger by the tail, as one Democrat, Thomas Rockwood, expressed. Driven to the point of exasperation by the ornery Day, Rockwood exploded. He had put up eight hundred dollars to "get the old man over here," but "damned if he wouldn't give twice that amount to send him back."

Still there was no better editor or newspaperman in the history of the San Juans. As John Turner, who worked for Day, recalled, Day was "big hearted," "a paragraph writer [without] peer," and a man who let the "chips fall where they may." Day liked his "plug tobacco and liquor." Turner provided the perfect epigram for Day's journalist efforts: his paper was "condemned by many, but was read by all."

Day, like Twain, made his paper lively. A few excerpts from the *Muldoon* and as it was renamed, the *Durango Democrat*, illustrate that fact.

> *Several plates of ice-cream*
> *And a piece of cake*
> *Make the finest kind of*
> *Modern stomach ache.*
>   November 5, 1880

> *Silverton has a boom. We saw it going in a few days since,*
> *three gamblers, two women and a yeller dog.*
>   April 27, 1883

> *The doctors have given up all hopes of saving the year*
> *1893 and death is expected inside of three days.*
>   December 29, 1893

> *The poorest man the ground above,*
> *is he who lacks a woman's love.*
>   February 15, 1907

Politically Day jabbed everyone from local to national politicians, but especially Republicans. Democrats loved him in Durango and elsewhere, although, as mentioned, he raised hackles among them as well.

No San Juan newspaper or editor outdid the Solid Muldoon *and its editor Dave Day for colorful, outspoken articles and reporting.*

> *He stood before the altar,*
> *Leading the Sunday school choir*
> *After he'd sold his proxy for $25,*
> *If he didn't, I'm a lyre.*
>   September 5, 1879

Political satire was one of Day's strengths and favorite weapons.

Dave Day's longevity continued into the twentieth century with his *Durango Democrat*, something that proved unusual. Another excellent newspaperman displayed a more typical career story: the ever on-the-go Gideon "Gid" Propper.

Propper started with Silverton's *La Plata Miner* in 1879–80. He moonlighted by writing San Juan articles for the national *Mining Record* and then became editor of the *Animas Forks Pioneer*. The excitement about the camp having a paper was shown by the fact that the first printed copy sold for twenty-five dollars and the second for eleven, the proceeds of which Gid donated to the school fund. Sadly, however, little Animas Forks could hardly support a newspaper. It folded, and the school did not fare much better.

Amazingly, he always landed on his feet. Propper edited the *Mining News* at Telluride, followed by Lake City's *Silver World*, then the *Red Mountain Journal*, and somehow managed to squeeze in a little time with the *Silverton Democrat* and *Silverton Herald*.

It was not that Propper was a poor newspaperman, quite the contrary. A rival wondered aloud, "How do you hook on Gid?" yet gave him a tip of the hat for "rustling more mining news" than anyone else. Propper was also hailed for creating "a spicy paper." Unfortunately the reasons for his travels, whether resulting from an incurable case of wanderlust or simply an inability to get along with owners, have been lost to history. So, sadly, have most of his writings. Yet he had a way with words: "Colorado without whisky would be like hell without brimstone."

The San Juan newspaper world was not solely the domain of men, as Caroline Romney proved with her *Durango Record*. The energetic, petite widow arrived in December 1880 to start the town's first newspaper. In slightly less than two years there, she helped shape the community in a way few others have equaled. A reporter later described her thus: "She talked with serene and animated grace that was simply charming." The gentility of that statement should not be misinterpreted. Like Day, her sharp pen jabbed at anything and everything.

She called southwestern Colorado "a land not only 'flowing with milk and honey,' but seamed with silver and gold and floored with coal." Her readers wanted to hear such flowery affirmations. Some perhaps did not want to read

Among the reforms editors advocated were crucial ones dealing with sanitation and sewage—the all-American outhouse, cold in the winter, hot in the summer, and smelly year-round.

---

about her push for women's rights, but the abundance of young men probably supported her plea that "we want girls! Girls who can get themselves up in good shape to go to a dance." She crusaded against Chinese "opium dives" and too many "worthless curs [dogs] about this city."

Five little newspapers were hissing and clawing like sassy kittens in Durango during Romney's stay. The competition was brutal to obtain readers, advertising, and matchless news stories. Outspoken, feisty, and colorful, Romney more than held her own, but Durango, like other San Juan communities, could not support such a large number of papers. A decline proved inevitable, and she left, along with most of her early competitors.

Whether Romney, Day, or anyone else, the first thing editors and their papers had to do was tell the world what they intended to accomplish. Rico's *Dolores News* (1879) promised to "be a mining journal in all that goes to make up a newspaper pertaining to giving the latest and freshest news about all mines." It would be "independent and candid in expression of opinions upon all subjects." And finally, the paper would be "the medium by which the world would learn of the existence, growth and prosperity of Rico and the pioneer mining district."

The *Creede Candle*, in 1892, proclaimed that "this is going to be a regular electric light. If you are a miner or in any way interested in mining you will want the *Candle*."

Then it promoted its new home. "You can't miss if you come to Creede. There is but one Creede camp and it is the richest mineral section opened since Leadville. Keep that in mind." The editor concluded, "Creede camp is a baby in years but a giant in stature."

Durango's *Southwest* pledged modestly, in 1882, to "pay its bills and undertake to collect its dues. It is a newspaper which is constituted as a business, and this business will be performed in all its branches, in a business [like] way." Finally, it proclaimed to be "a good advertising medium and will endeavor to be a first-class newspaper."

The *Candle*, the *Southwest*, and the *News* were off and running as were their competitors and contemporaries. None had time to take a second breath in this journalistic dog-eat-dog world. They would all try to do the things they promised and more while boosting their towns, mines, businesses, people, and anything else the editors wanted to promote.

H. L. Mencken described San Juan newspapers perfectly when he wrote, "All successful newspapers are ceaselessly querulous and bellicose."

Like their neighbors, newspapers started off under primitive conditions. Caroline Romney printed her first paper in a tent in the cold of December 1880. Still, she told her enthusiastic *Record* readers in flowery Victorian prose, she was pleased to be in "this city in the wilderness, the Mecca of our ended pilgrimage." They were as happy to have her in town as she was pleased to be there.

With less fanfare the proprietor of Rico's *News* purchased a lot, found "progress slow" on building an office, and thus also secured a tent. With "no desk, chairs or tables" and using the ground for a table and a candle box for an easy chair, he published his first issue.

Typically, the newspapers were four pages long. Boiler-plate, or ready-print, articles or news items, clipped from other papers or the news wires, took up most of the space, with a page set aside for local and other state news. Advertisements covered as much space as could be sold and included the ever-popular and profitable patent medicines and state and local ads.

Locals desired a lively sheet, and most editors strove to achieve that goal. None was better at doing so than Dave Day. Even his rivals admitted this amid their jealousy and bickering. The *Solid Muldoon*, acknowledged the *Dolores News* (September 23, 1879), "pops the individual who takes it up and opens its folds, like a little shock of the electric battery." In a writing style that would have made Day proud, the editor continued. "It sticks to the mind, indelible and fixed like plaster to a poor man's ribs." Most papers were less kind to Day, as he was to them.

Day and other editors were expected to boom the local community in every way possible. The *Ouray Herald* (June 22, 1899) typified this expectation. "The Gem of the Rockies," as it referred to Ouray, "is without peer among towns and cities of the world for the magnificence of its surroundings, the grandeur of scenery and the rare combination of low latitude balanced by high altitude." All this, in the editor's eyes, produced "equable climatic conditions." Ouray offered everything a tourist—and Ourayites and others were pushing tourism—and potential resident could want.

A local paper was expected to defend every corner of the community from jealous rivals. The *Silver World* in April 1876 started with this boast: "As a town we have not now nor will ever have a successful rival." The towns along the Animas "have long since ceased to have hope," so move over and "give Lake City its due." The comments fell on deaf ears. Others would not cheerfully move over, nor would they do so later.

There were ongoing newspaper feuds between almost all the towns. While some were in fun, or seemed to be, behind them lurked a deadly seriousness. A few examples will indicate the combat.

The streets of Silverton present a lively appearance now-a-days, while poor Ouray still howls for a railroad.
   San Juan, April 21, 1887

The polar camps of Lake City and Silverton must get their cereals and vegetables from the Uncompahgre valley. Neither is blessed with enough agricultural land on which to set a hen.
   Solid Muldoon, March 12, 1880

Half of Hudson Town [better known as Red Mountain] is sporting women and their pimps. Red Mountain City supplies all its small neighboring camps—Hudson Town, Rogerville and Ironton—with meat and potatoes.
   Red Mountain Pilot, May 5, 1883

At the same time, if any foolish, or jealous, outside paper criticized the San Juans, the local press sprang to the defense. Denver, particularly, appeared to frequently antagonize local editors. Internal brawls were acceptable, but outsiders could not criticize the region without several papers firing back.

Money kept the papers going, and money was sometimes hard to find or collect. The *Telluride Republican* weighed in on July 3, 1886, with these admonitions:

"The value and usefulness of a newspaper depends largely upon the support it receives." The businessmen have "generously" supported the paper with "healthy advertising and patronage. Now it is up to the miners and citizens generally to accord a good subscription list."

A year earlier one of its rivals, the *Evening Journal*, had philosophically stated: "We will never kick if we go broke on account of lack of patronage. It is the duty of a newspaperman to keep the business going." The editor did wish that "the belly-aching element of all mining camps could be induced to give up a dollar once in a while." They must not have because the paper was gone within a year.

Little sympathy existed for a fallen rival. Silverton's *La Plata Miner* (January 17, 1885) asserted about its neighbor the *San Juan Herald* that "it is rapidly approaching dissolution," then added that "the sooner it ends its miserable existence the better. There will be no mourners to attend is funeral." The editor's prediction proved correct; within a few months the *Herald* ceased its "miserable existence."

Occasionally, a paper confessed that it had trouble finding news. It was "difficult to print a newspaper of any class or kind in Rico today," bemoaned

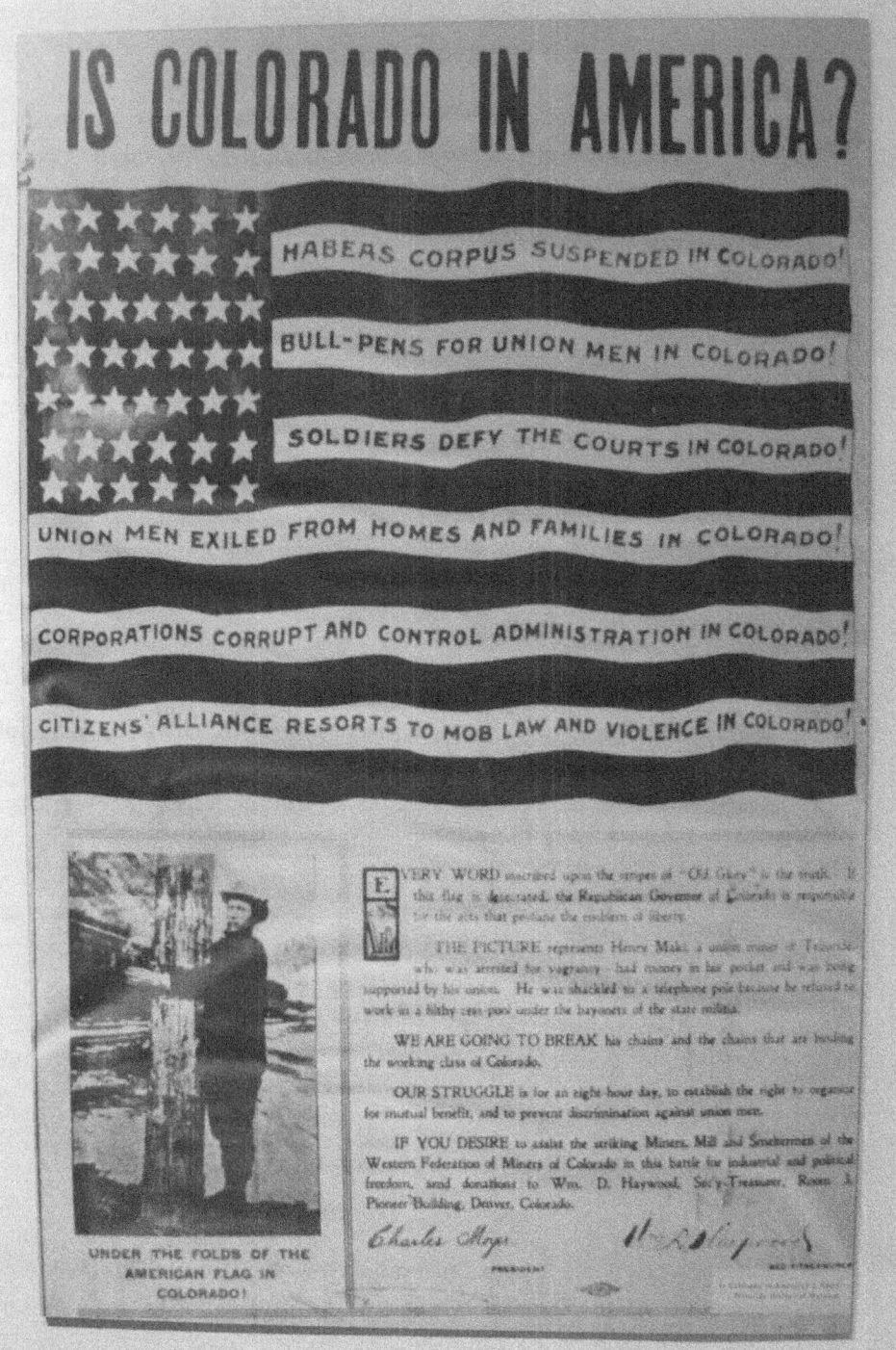

Labor and management stood eyeball to eyeball in 1903–4—a one-sided fight, as San Juan miners found out. Owners held the power.

Raising Hell and a Lot of Other Things

the *Dolores News* in January 1880. The editor put his finger on the reason: "Progress on development cannot well be reported while so much snow lies on the ground."

Sometimes a paper tried to tell its readers that it could not boom the town alone, as the *Ouray Herald* did on January 2, 1896, amid the gloom of the 1890s depression: "Now that we have entered upon a new year we most respectfully say to our citizens not to depend upon local papers to boom the town and bring to it prosperity." The paper would do all "that can be done," yet the press "expected citizens to put their hands to the wheel and help along with the good work." All Ouray lacked, the *Herald* contended, "is enterprise of those interested to restore the busy scenes of five years ago."

A couple of weeks later the *Herald* took local "merchants, professional and businessmen" to task for not advertising in the paper. This theme had been a common one in the San Juans for years. "It is extremely aggravating to hear men kick about local papers who never spend a cent in advertising. The same men are always finding fault about dull times." The paper pointed out that "prosperity of your local papers means your prosperity. Give us a chance."

After the turn of the century, local papers carried many more national ads, probably because of the income generated and the fact that they arrived ready to publish. Telluride's *San Miguel Examiner* (April 27, 1907) exemplified this trend, offering a wide variety of ads for its readers. Among the advertisements it carried were those for the International Correspondence School, Carter's Little Liver Pills, joining the navy, Lydia E. Pinkham's Vegetable Compound, and Cuticure soap and ointment. If one believed all she or he read, these products promised to cure headaches, colds, coughs, rheumatism, piles, and a host of other problems. Both men and women would benefit if they believed.

The *Examiner* was one of the largest daily papers in the San Juans, with ten pages; local news and happenings appeared on half those pages. News taken from the wire services filled many columns. Major league baseball results were included. The paper still promoted the local mines, but no editorial attacks on neighboring districts or camps sallied forth. Those were saved for currently booming Nevada.

The paper was one of the rare ones that endured—it turned twenty-two in 1908, although it had started under another name. Throughout those years it had "done all in its power to build up the main interests of the town and county." The *Examiner* proudly boasted that its "circulation is larger now than ever before." In addition, it carried news its readers of twenty years before could not have imagined—news about the automobile, movies, and the flying machine. Some San Juan papers even printed photographs.

Most of its earlier contemporaries had long since folded and departed. A few hung on as weekly papers. The newspaper business never proved as easy or as profitable as its backers had hoped. Many of the small camps the papers once boosted were gone too, along with many customers from the larger towns. Only Telluride, Silverton, and Ouray still had any major mining, and it had become corporate controlled. The booming excitement of earlier years, which the papers had chronicled, was gone as well.

In many respects, by World War I the remaining San Juan papers looked much like other small, non-mining-town newspapers. Life had settled down and become routine, and only a small or occasional mining column reminded readers of yesterday's heritage.

Over several previous generations, San Juan newspapers had done yeoman's work—promoting, defending, reporting about, and occasionally chastising their communities and neighbors. They had lived up to the newspaper's duty "to print the news," and more than a few of them had "raised hell."

Railroads were the key to the San Juans.

CHAPTER FOUR

Transportation Revolution

*The Stub, a small dingy train, was to carry us to Telluride.
The engine, baggage car, and dismal passenger car were relics of
a past generation. With great effort the small locomotive shuddered
and jerked into motion on a narrow, shay roadbed.*
—Harriet Backus, *Tomboy Bride*

*One begins to curse the roads immediately after he leaves Antelope Park,
and, though rough roads have been experienced before, at the summit of
the range roads that are—not roads at all, are encountered.*
—W. B. Dickinson, November 1, 1874

From the 1870s into the second decade of the twentieth century, the San Juans were part of a transportation revolution. In 1870, travel occurred by foot, wagon (with a variation of the stagecoach), or animal, just as it had for centuries. Indeed, it would have taken a Roman centurion, or senator, from the time of Caesar little time to adjust to travel in southwestern Colorado.

By 1914, the railroad had been chugging in and out of and around these high mountains for a generation. The automobile had arrived, and an aeroplane was not something folks read about and wondered what the world was coming to. One had actually taken off in Durango, briefly setting the world's high-altitude record.

All this remained in the future, however, for the pioneers who walked or rode into the rugged San Juans as the 1870s opened, and permanent settlement lay just around the corner. By circuitous routes wagons could reach Animas City, next to Durango's location a decade later and to the cabins that

The town without a railroad was the town without much future. The shout of "Three cheers and a tiger" echoed around town when those iron rails arrived.

would someday become Ouray and Silverton. In all these trips, however, they had to trespass through land guaranteed to the Utes by treaty.

The future site of Lake City proved the easiest to reach. Beyond it, though, lay the Engineer and Cinnamon passes, which had to be crossed to travel into the heart of the mountains. Both presented barriers well over twelve thousand feet and were crossed only by trails little improved over what might be defined as game trails.

Those early trails, really only rough paths, into the San Juans were dangerous. William Henry Jackson ventured in with the Hayden Survey in 1874 over Stony Pass, the primary route into Howardsville, Silverton, and Baker's Park. He exclaimed, "What can possess those people we pass to go into that place this time of year [September]. Passed burro trains & wagons. Smashed up [our] wagon there," adding that it was a "poor excuse" for a road.

Alfred Camp struggled over the same trail and described his feelings: "This experience will hardly be forgotten as for several hours we scrambled and slid as best we could to get down—fortunately without breaking our necks." He continued, bitterly describing Stony Pass as "almost an impassable road winding upwards. At present it is the only route over which loaded wagons can be brought in. In several places we noticed where wagons had to be 'snubbed' down by ropes or chains. It is as hard to get an empty wagon out as to bring a loaded one in."

No matter which way one traveled, the route remained seasonal. As soon as the snow melted, prospectors wandered in; when the winter storms arrived, they left unless they were extremely brave and well prepared or foolish. Then, as much as six months or longer, snow and ice curtailed, if not stopped, travel.

One *could* get around on snowshoes, as they called skis in those days. Only a few people had even seen them, though, much less mastered the techniques needed to use them. There are a few stories of skis being used in the nineteenth-century San Juans, but except for recreation they were not practical for getting supplies, beyond mail, into those mountains.

The best and sometimes almost the only way to do so involved using trains of pack mules and burros. Animals hauled nearly all goods brought in and the ore moving from mine to mill in the early days. The surefooted burros could live off the land, but mules had to be fed, and neither could carry a heavy load—often just over two hundred or so pounds for the mules, and the smaller burros much less—uphill. Most of the San Juan trails involved much uphill climbing, which cut down on the freight the animals could carry. Looking over the San Juans, some people swore it was all "up" in those mountains.

Without better transportation, even if not year-round, the cost of living would remain high, mining stymied, the little camps stagnant, visitors and investors generally uninterested, and the future of the San Juans in doubt. The highest mining region in the United States faced dismal prospects without improved transportation.

This became a continual theme in San Juan newspapers and among locals in the 1870s. Said one 1874er, as quoted earlier, the roads were "the roughest imaginable." W. B. Dickinson continued in a letter to Denver's *Rocky Mountain News*. "One begins to curse the roads immediately." Curse they did, but it seemed the early San Juaners could do little else.

The *Ouray Times* (July 14, 1877) rated building roads as one of the most significant reasons for making Ouray "second to no town in this portion of Colorado." The paper pointed out that the trail to Mineral Point "is sadly in need of repair and in some places almost impassable." It was in Ouray's interest to "make it as good as circumstances allow." One Ouray resident testified in court that a man "risked his life on [that trail] and it was not safe to ride a horse over it."

Lake City agreed. The *Silver World* (April 20, 1878) declared, "Lake City must employ its energies" to open "communication with more distant and inaccessible camps by good wagon roads." If it did not, the community "must subside into a local mining town unless railroad enterprise particularly favors her."

Two events improved the opportunities for better transportation. First, the Utes ceded the San Juans in the Brunot Agreement in 1873. Misunderstandings, cultural clashes, distrust, and Nathan Meeker's troubles with the White River Utes, all of which led to the Ute War of 1879, finally solved that problem. All but the Southern Utes, in the southwest corner of Colorado, were removed by 1881.

Meanwhile, the 1874 creation of La Plata County, and the splitting off of San Juan County two years later, created organized government. Hinsdale and Rio Grande counties appeared that year, and Ouray over took the northern western part of San Juan County the next year. Finally, in 1881 the southwestern part of Ouray became Dolores County, then San Miguel was carved out of the western part of rapidly shrinking Ouray. Except for some minor tuning, the San Juan counties were formed, with Mineral County joining their numbers in 1893.

County structure allowed roads to be surveyed, built, maintained, and improved. Further, counties could give charters for toll roads. A road system started to appear almost immediately along paths beaten down for centuries by animals, then Utes, and finally prospectors and miners in more recent times. Of course, all this was easier said than done, and the expense and tribulations of developing roads nearly overwhelmed the young counties.

One answer was to have entrepreneurs build toll roads, although it was not always the best solution, complained San Juaners who had to pay the tolls. Russian immigrant and Saguache merchant Otto Mears, who provided funds and hired crews, filled that gap and was hailed as the "toll road king of the San Juans."

Mears's first road went from Saguache to Lake City, then branched up Henson Creek over Engineer Mountain to Animas Forks. Eventually he rebuilt the road to Silverton. In 1877, Mears opened a road that swept northwestward in an arch to reach Ouray. In time, further toll roads connected Dallas and Telluride, Durango and Parrott City, Ouray up the steep and narrow Uncompahgre Canyon to Ironton Park, and Silverton to Red Mountain.

Mears was not alone. James Wightman and partners built a toll road that went southward to Animas City in 1877. Silverton now had outlets in two directions. A few short roads, such as one from Del Norte to Lake City and another from Rockwood to Rico, completed the system. Others were enthusiastically planned but remained on paper.

Complaints soon arose. The roads were not maintained, they were too rough, and tolls were too high. The public quickly thought they should be county roads, free of tolls. Eventually that was what happened, or else the roads were abandoned as traffic and profits shifted elsewhere. Otto Mears received the brunt of these complaints, but he also deserved praise, as the *Ouray Times* proclaimed in 1877. "Mears is doing good work for this section of Colorado. The roads are certainly very much needed." The *La Plata Miner* (December 1884) also came to his defense. When the counties lacked the means to build roads, commendably Mears had built toll roads. A "toll road is a long way better than a burro trail," the paper added.

Winter weather caused grief for all types of transportation, and the San Juans were blessed with an abundance of "the beautiful white stuff" as more than one frustrated San Juaner expressed. Not only did snow close the passes, but freighters also tried to get their pack animals into the lower valleys for winter pasturage. Customers faced costs "three to four times the summer rates," if one "is fortunate enough to secure transportation of goods at any figure."

Stagecoach service beyond Lake City and Ouray proved impractical because of the punishing mountain terrain and poor roads. Barlow & Sanderson, the leading express and stage company of the day, avoided those mountains. The company reached the outlying towns, but despite claiming that it served all areas of rapidly expanding 1870s Colorado, the San Juans remained on the sidelines.

Poor transportation caused many problems and inconveniences, perhaps none more annoying than mail delivery. In 1877, Ouray residents called the situation "shameful" and wanted more than weekly service. They thought it should be "tri-weekly at least." For some communities, such service came earlier. Lake City's *Silver World* cheered the arrival of triweekly service, which brought mail from Saguache to Lake City and "thence to Silverton," weather permitting, of course. Ease of access made this service possible, but the editor correctly pointed out that a "lack of mail facilities has been one of our most aggravating evils."

Despite promises of mail-coach service, that service still remained off in the future. Winter snows forced carriers on skis to deliver mail. Three carriers made the run from Lake City to Silverton in relays in the winter of 1875–76. The last one covered the distance between Animas Forks and Silverton. With blizzards and snowslides, the men took their lives in their hands to get the mail to post offices. The mail might have gotten through, but this was stopgap service at best.

By the late 1870s, despite this multitude of troubles and complaints, the San Juans featured a network of trails and roads that were good enough,

The caboose served a variety of purposes, including occasionally carrying a few passengers.

for instance, to allow operation of stagecoach lines. Barlow & Sanderson reached brand-new Durango in 1880 and faced a rival in the local Pioneer Stage Line that ran stages to Silverton, Parrott City, and Rico among other places. Colorado's silver-mining entrepreneur purchased the firm and renamed it the H. A. W. Tabor Pioneer Stage and Express Line. Even after the arrival of the railroad, it ran from the Silverton line to Rico, which did not yet have train connections.

Easterners might have been shocked at what locals called roads that these stages bumped and swayed along over, but the situation had definitely improved. Indeed, when Rico burst on the scene in 1879, it easily got a road to the south and into the Montezuma Valley, as well as one to Silverton and soon northward to the future Telluride.

Still, in spite of such progress, the great hope rested with that nineteenth-century wonder, the train. Almost as soon as the San Juans were revealed to have mineral potential, San Juaners called for railroad connections—something they fervently believed would be the answer to all their transportation problems.

The answer to those dreams lay with the Denver & Rio Grande Railroad (D&RG). Coloradans had been railroad conscious ever since the territory had been virtually bypassed by the transcontinental railroad, except for Julesburg, while its neighboring rival, the future Wyoming, sat astride the main line. Stubbornly refusing to give up, Denver's leaders built the Denver Pacific from the territorial capital northward to intersect the Union Pacific. That was quickly followed by the Kansas Pacific, with both arriving in Denver in 1870.

William Jackson Palmer, in the meantime, developed the idea of the D&RG running north and south, tapping natural resources all the way into Mexico. He never got that far. After losing a battle with the Santa Fe Railroad over the right-of-way over Raton Pass and New Mexico beyond, and being lured into the mountains by the Leadville silver excitement, the D&RG became a mountain railroad.

Palmer looked longingly at the San Juans as a potentially profitable market, and his narrow-gauge cars (three feet between the rails) were perfectly suited to tackle the high valleys and mountains. Narrow-gauge railroads could climb steeper grades, go around sharper curves, and were cheaper to build. The trade-off was that the railroad had smaller engines and cars. Nonetheless, Animas City in the valley and the mining communities in the mountains eagerly awaited his arrival.

Plans were launched in 1879, and surveyors soon arrived on the scene. Meanwhile, Palmer did what he had done along all his lines: he asked communities his railroad encountered to help with expenses to ensure they achieved the desperately needed connections. The leaders of Animas City refused to heed history and would not negotiate terms. The first impact of the D&RG in the San Juans was the creation of Durango in September 1880, two miles south of Animas City, which had doomed itself.

Meanwhile, Silverton was beside itself with anticipation. Silverton's *La Plata Miner* watched developments with keen interest. It offered no sympathy to Animas City. The "beautifully located" new town "will knock the stuffing out of the present town, yet it will be a good thing for us all, and especially our San Juan neighbors," the paper gushed. By November 1879, a surveying party was camped right below Silverton after "locating the line through Animas canyon," a "slow difficult task," the paper concluded. It also proved to be "attended with some danger to the surveyors who have been let down ropes over canyon walls in order to get level" road grades.

Many a passenger spent a long time waiting on these benches. Dave Day thought the 1907 dilapidated depot ought to receive a "pension from the Cliff Dwellers' League."

In an article headlined "Look Out for the Locomotive," the *Miner*, on November 29, 1879, showed its excitement. Contracts had been let and were to be completed by July 1, 1880, to the lower Animas Valley. In the paper's view, the once-rough, narrow road to Animas City had become a "magnificent wagon road, always open." Surely, it would not take the D&RG long to go up that route to Silverton.

This would benefit the town in every way, from shipping ore out and goods in to reducing rates. It would bring ease and comfort, as well as investors with money in their pockets. "In fact, it is impossible to estimate the great advantage in every way completion of this road will be to our camp," the paper proclaimed. On December 20, 1879, an editorial predicted with the railroad's arrival "perhaps next year, then we will see the beginning of a boom for this country that will not cease growing for a hundred years to come."

The year 1880 arrived but not the D&RG, not even into the Animas Valley, except for its planned new metropolis, Durango. Silverton never gave up hope and jumped on any rumor that surfaced, including one that purported that four hundred men were working on the grade below Silverton in late 1880 or one that "railroad men who are supposed to know" said the line would reach Silverton by "January 1, 1882." Wrong again.

The year 1881 finally saw progress. The tracks reached Durango in July, and the first official train arrived early the next month. Construction then moved northward along the surveyed grade. The *Miner*, though, was beside itself. The slowness, high hopes, and desperate need had all taken a toll. When it was learned that some of the leaders of the D&RG planned to build a smelter in Durango and purchased Silverton's and tore it down, it proved too much. Palmer and his cohorts, the paper charged, "want to buy mines at 5% of their real value," which was why they destroyed "the most successful smelting works in all of southern Colorado." That comment represented local pride and anguish; in reality, the smelter had never been successful.

The editor asked, "Does anyone know of a prosperous town or community that existed in Colorado before the construction of the Denver & Rio Grande which the railroad has not tried to kill and destroy?" Wound-up, the editor let loose a final blast: Palmer and his "skinners" would do "everything in their power to make Silverton valueless."

Yet by Christmas 1881 the *Miner*, perhaps caught up in the season, became more charitable toward the oncoming iron horse. The arrival would create

When the Denver & Rio Grande Railroad arrived, both mining and communities boomed as never before. It also brought the tourist that came to see the "elephant."

a "mining revolution," it proclaimed, and "make Silverton a mining town second to none in the state."

The *Miner*'s rival, the newly arrived *San Juan Herald*, was much more understanding. It cheered when the Silverton branch was opened to Rockwood in January 1882, shortening the stage trip, via Tabor's line, to Silverton. A passenger could now leave Silverton at eight in the morning, make the train from Rockwood to Durango at two thirty and arrive there at half past three—an unheard-of pace. Rockwood promptly enjoyed its moment in the sun with a brief, brief boom.

As the months raced by, although hardly fast enough for Silverton, more men were put to work, grading was completed, and bridges were built. Now the line was forecast to be completed by April 20, 1882, but that was wrong again. Then the *Herald* forecast a mid-June arrival with these encouraging words: "The difficult problem of our prosperity will then be solved, and we will then have to thank the enterprise and pluck of the much maligned narrow gauge."

Dave Day thought Durango's depot so dilapidated that "cockroaches and bed bugs played baseball on the depot's ground floor." "Entertainment" while waiting perhaps!

The railroad was already advertising the tourist possibilities along its route. These included hot springs for invalids, cliff dwellings for the adventuresome, the leading mountain resorts, and Royal Gorge for tourists. Silverton did not care about the promotions; it only wanted rail connections.

Finally, on June 27, 1882, the whistle "of a D&RG engine was distinctly heard in Silverton." Not until mid-July, however would a train appear—late once again—this time for a big July Fourth celebration. It did not matter; the special holiday train from Durango was met outside the town by carriages to bring in the Fort Lewis baseball team and band, as well as a train filled with fans and dignitaries. Crowed the *Herald* in its July 6 issue, Silverton sat on the "threshold of long expected prosperity." The paper joyfully noted a week later that "we will help the railroad and the railroad will help us. That is about the size of it."

Even the *Miner* joined in. "So far, all that can be done by the outside world has been done . . . by this medium it has . . . opened to us—what now remains is for us to do—to commence to make ourselves and make good on our statements." The editor had seen "a couple of lady fortune tellers perambulating on the streets." He could not resist a warning: "Silverton heretofore has been exempt from these parasites [including beggars of the professional class] and other groups." The "advent of the railroad may be expected to inflict [them] on us in increasing numbers."

In the years that followed, Silverton's railroad cup overflowed. Thanks to Mears, the Silverton Railroad reached Red Mountain and Ironton in 1888, and the Silverton Northern eventually reached Animas Forks in 1904. In time, the Silverton, Gladstone, and Northerly (1899) also fell under his control. Frustrated at being unable to go down the steep and narrow Uncompahgre Canyon to Ouray, Mears completed his railroad empire by building the Rio Grande Southern from Durango to Ridgway, with a branch into Telluride, in 1890–91.

The Denver & Rio Grande, meantime, extended lines into Ouray in 1887 and Lake City in 1889. When Creede boomed in the 1890s, the D&RG purchased a line to the camp "that is now all the rage and is certainly wonderful." Across the San Juans the railroad's coming electrified mining interests and energized towns, drowning out pessimists' protests.

Some of the protests warranted an audience. The *Red Mountain Pilot* charged in 1883 that the D&RG had discriminatory rates for shipping ore that favored Pueblo and Denver smelters over Durango's. Dave Day believed ore-shipping rates were "unjust" and they forced a "number of mines" to close down. Others complained about "excessive" passenger and freight rates, and some wanted better time schedules and on-time arrivals of trains.

For the railroad traveler, a fine meal and a good night's rest awaited at Ouray's Beaumont Hotel. The sheets were as clean as the "wind driven snow."

All these complaints aside, San Juaners were more than pleased to finally have their railroad connections. Towns and camps now became railheads, and from there wagons and pack trains transported goods higher into the mountains to more isolated spots.

Despite harsh winter weather, featuring snow for five or six months, the railroads kept going. Sometimes slides shut them down, or lack of traffic did not warrant clearing the tracks until spring arrived. The decline in the price of silver and the closing of mines eventually hurt the little lines out of Silverton that depended almost entirely on mining business.

By the 1900s, railroads were searching for new revenue sources, and tourism popped into view. Mears thought Silverton would be an excellent summer resort, with its "delightful summer climate" and its "grand scenery." More practically, the D&RG pushed the circle route from Denver to Durango to Silverton to Ironton, then by stage to Ouray, and back to Denver by train. A side trip over the Rio Grande Southern could be taken to fascinating Mesa Verde, which became a national park in 1906.

Silverton was not alone in luring tourists. "Peerless Ouray, Most Picturesque Town in the World," Creede, Durango, and others joined in the rush. On came the tourists to see the sights, natural and otherwise, and savor a fast-vanishing West. Times were changing; the railroads now pushed tourism rather than mining. Then the automobile chugged onto the scene. The first car into Silverton, after struggling over old Stony Pass, aided by a team of horses, arrived in August 1910. Then, again assisted by horses, it ventured on to Ouray. The next year, the first San Juan County resident purchased a car, and local horses had to be on the alert for the new, backfiring, hissing monster.

The call went out promptly for "better roads." It was exciting news in Creede, in August 1914, when five "automobile loads of tourists" drove into town "direct from Kansas City." Their only problem had been muddy roads; that notwithstanding, they "liberally" praised Mineral County roads. Then, with the airplane soaring in the skies, old-timers of a generation before, who had labored across Stony Pass on their way to mine the gold- and silver-ribboned San Juans, stood amazed at the transformation to their "Silvery San Juans." They had seen a transportation revolution before their eyes.

Being unable to shop daily, the housewife purchased large sacks of flour, sugar, and other items. In the mountains, the purchase of a hundred-pound sack was not uncommon.

CHAPTER FIVE

"Love Can't Live on Heavy Bread"

Then, proudly, I handed him the platter with the beautifully browned mutton, so crispy looking. . . . George was looking not at the roast but straight at me with an expression of pity, embarrassment, and what shall I do now. "Dearie," he said in his gentle manner, "I am afraid this meat did not get quite done." It was still frozen solid.
—Harriet Backus, *Tomboy Bride*

Patrons of the telephone are requested not to hold the line open talking nonsense for an hour, especially when near meal time. A housewife is sometimes very anxious to get the attention of her grocer or market man.
—*Ouray Plaindealer*, January 20, 1905

he English poet Lord Byron wrote:

All human history attests
That happiness for man—the hungry sinner!—
Since Eve ate apples, much depends on dinner.

If Byron was right, and surely he was, then much depended on the skill of San Juan housewives. They faced many problems in preparing that dinner—altitude, limited selection of cuisine, and the high cost of goods at the local store. The same might be said for the bachelor miners, who likely consumed the traditional coffee, beans, and whatever meat might be available.

Well-cooked, wholesome food proved one of the major keys to survival in the San Juans during the mining era. Hard work, the order of the day, demanded the best possible meals, as did one's health. As a result, the cook's skill in purchasing and preparing the provisions for meals often meant the difference between a healthy life-style and a host of problems.

The *Golden Manual, or the Royal Road To Success* (1891) advised the newly married housewife on how to maintain domestic happiness:

> We [men] very soon get tired of heavy or burnt bread, and of spoiled joints of meat; we bear them for a time, or for two perhaps but about the third time, we lament inwardly; about the fifth time it must be an extraordinary honeymoon that will keep us from complaints.

What would the result be? Soon, the "fire of love [was] damped," and husbands find "too late, that we have not got a helpmate, but a burden." The roles of husband and wife were definitely well defined in Victorian America. One symbolized the queen of the kitchen, the other the breadwinner.

The wife might be queen of the kitchen, but cooking at high elevations in the San Juans was not like cooking at sea level. Harriet Backus found this out as a young bride from California when she journeyed to the Tomboy Mine at eleven thousand feet above sea level. With brutal honesty she admitted, "I was one bride who couldn't boil an egg."

She confessed to some of her failures, including her attempts at making bread. When she tried for the second time to make dough, Harriet was pleasantly surprised after leaving it overnight to rise. "I was delighted to find the dough near the top of the bucket. I molded it into loaves, let them stand the required time and baked them. The bread came out golden brown, tempting, the wonderful aroma of freshly baked bread throughout the house." Alas, she cut a warm slice. "What a disappointment! I almost cried. The loaf was nothing but a mass of holes with a webbing of dough, resembling genuine Swiss cheese." Her tribulations had only started:

> Only after repeated trials were our frozen eggs boiled long enough to be palatable. It was hard for me to realize that water boiled at only 190 degrees and to determine the additional time required to boil an egg.
>
> At home in California I had made delicious cakes and decided to use one of my mother's recipes. I mixed the batter with great care and put it in the oven for the required time to bake. The result—it remained batter.

Every home had its kerosene lamps, washbasin, and water pitcher. A mirror helped in a variety of ways, from shaving to getting ready for the dance

I tried a pot roast and we eagerly anticipated dinner. It was browning beautifully and a savory odor came from the pot. No knife we possessed would cut that roast.

One day I said, "All miners eat beans. Do you like beans, George?"

"Yes, of course," he answered, always amiable about my experiments and probably hopeful of a triumph. So for two whole days I boiled beans. They neither swelled nor softened, but remained as hard as marbles.

Harriet persevered. She spent many a day perusing the *Rocky Mountain Cook Book* and mixing recipes. Finally, she reported, "I learned to cook the hard way, but it took time." Fortunately, she had a saint for a husband. "Yet, never during all the years did George ever complain or criticize my efforts."

Other wives were much more experienced at cooking at high elevations, but even the best of them faced similar trials and troubles. They not only had to deal with the elevation (the highest mining district in the United States) and the weather, but they also had a limited market from which to choose supplies for their meals. They had to learn, as Harriet did, to order for a month or better to be supplied for the shortages that were bound to come. Winter was particularly a trial, with snow-closed roads and slides that stopped even the most powerful railroad engine, resulting in shortages on store shelves and in pantry cupboards.

The miner's cabin probably had a fireplace for heating, but the potbellied stove proved more efficient. Both, however, provided comfort on a long San Juan winter evening.

The high elevations of the San Juans did make storing perishable items easy during much of the year. One old-timer grumbled as he tramped out of the mountains that he would not work in a place with "three months of mighty late fall and nine months of damn hard winter." At an average elevation of well over nine thousand feet, perishable goods could easily be kept frozen except in the high summer months, and even then it was little problem to keep them cool. Rotten food came in the frozen form, however, much to a cook's dismay.

Many a fine dress or shirt came from this machine, not to mention repair jobs by the dozens. Alas, the bachelor hoped to find a seamstress to make needed repairs.

It was more than just time-consuming and frustrating to experiment with cooking; it was also costly because of the prices of food and everything else. That was a common misconception about moving west, the high cost of living, something guidebooks trying to lure people often ignored or rationalized away as best they could.

Two such guidebooks appeared in the 1870s, the *Williams Tourist Guide* and *Rocky Mountain Tourist*. They did their best to convince their readers that such problems could be overcome or, at the least, minimized. The former argued that prices were only "a little higher" than those in the plains cities and that was because of freight charges. The latter encouragingly told its readers that "ruling prices for the substantials of life are by no means as high as many people might imagine."

Both guidebooks wisely recommended that it was best to store up in the summer for winter shortages, an excellent suggestion. They knew whence the problem came. "It is when the market is poorly supplied as in winter or early spring that prices are high. Then it is the merchant's harvest, for prices are only governed by his caprice or conscience."

A *New York Times* reporter noted (November 3, 1876) that prices in Lake City were not "so high as might be expected." The average prices, he thought, "are about 25% higher than in Denver." That did little for the miner or the worker whose wages were not 25 percent higher than those in Denver. The situation moved the *Silver World*'s editor to suggest (April 20, 1878) that wages be raised "to something near what they ought to be in this inclement and expensive region." Wages would not be raised in Lake City or elsewhere in the San Juans.

A sampling of prices in Rico in the fall of 1879 found beans at ten cents, potatoes at seven cents, flour at ten cents, and beefsteaks at ten to fifteen cents per pound. In Denver flour and potatoes cost four cents per pound, and other items were similarly lower. The Denver market also offered much more variety.

Times got better, and the cost of living decreased somewhat with the coming of the railroad in the 1880s. It also helped that farming started in the valleys to the north and south of the San Juan Mountains. Silverton's steaks now cost twelve cents a pound, potatoes were two cents a pound, and fresh eggs—a rarity in the 1870s—cost fifty cents per dozen.

By the 1890s, not even a new camp suffered through the usually high-priced early days. The *Rocky Mountain News* (January 19, 1892) noted that supplies in newly opening Creede cost "but little more than in the San Luis Valley. This state of affairs is rather unusual in a new camp which has a veritable boom." The fact that Creede was next to an agricultural region, with ease of transportation and communication, the paper decided, "accounts for it." While the "necessaries of life such as whisky are not much dearer" than in Denver, flour, bread, meat, and so forth "are 25% higher," a percentage no lower than earlier. Fortunately, Denver's prices had declined, producing a lower base to work from than in the 1870s.

By the late nineteenth century, merchants in the mining communities faced low-priced competition from the outside, mainly through the Sears and Wards catalogs, which offered a variety of goods no local store could come close to matching. As Harriet Backus described the familiar catalogs while looking for a "folding bathtub," "[I was] leafing through the mail order catalogue, the dog-eared standby of miners' families." She found one and was able to take "the time-hallowed Saturday night" bath, with "ladies . . . first even in the mines." Husband George used the same water, but a bit cooler and murkier. Heating water on the stove was burdensome and time-consuming, so sharing was not uncommon.

Few merchants likely carried folding bathtubs in stock, but that was not the real problem. A camp or town business could not match catalog prices, even factoring in the cost of shipping to the San Juans. Shipping one hundred pounds of goods from Sears, Roebuck in Chicago to Silverton or Ouray, for example, cost only about four dollars for first-class freight or two dollars for fourth class.

The 1897 *Sears, Roebuck and Co. Catalogue* proclaimed itself to be "the cheapest supply house on earth." The company was not shy about touting its pricing advantage. "If you buy from us you will get your goods for as little or less money than your local dealer does."

The Sears grocery department urged its readers to "BUY OUR GROCERIES AT WHOLESALE. In no other line of merchandise can [you] save so much money in a year. Your local dealer buys at wholesale and makes money on the groceries he sells you. YOU CAN BUY AT WHOLESALE PRICES and save this profit." The prices listed must have driven local merchants to distraction.

> Coffee 15–25¢ per pound
> Canned vegetables 9–13¢ per can
> Hams and bacon 6–13¢ per pound
> Dried fruit 5–13¢ per pound
> Sugar no price listed: "We guarantee for you the lowest
>   importation price on the day your order is received.
>   WE ASK NO PROFIT ON SUGAR."
> Flour 90¢ to $1.20 for a forty-nine-pound sack
> Beans 2–5¢ per pound

Outside competition cut into local business, and no merchant could match the prices just quoted and make much or any profit. Customers did not complain, but businessmen did. The *Lake City Times* (October 1, 1896) spoke for many merchants when it recommended that people buy at home and not from "foreign" places. Home merchants paid local taxes, supported local endeavors, and were the heart and soul of the community.

Rico's *News-Sun* also complained, saying that the "people of Rico have been the greatest enemies toward the advancement of our own welfare." It seemed that in the heated political free-silver atmosphere of the 1890s, Populists, Republicans, and Democrats boycotted each other, as did some merchants who "purchased goods on the outside they could have obtained at home." This led the miners to follow their example.

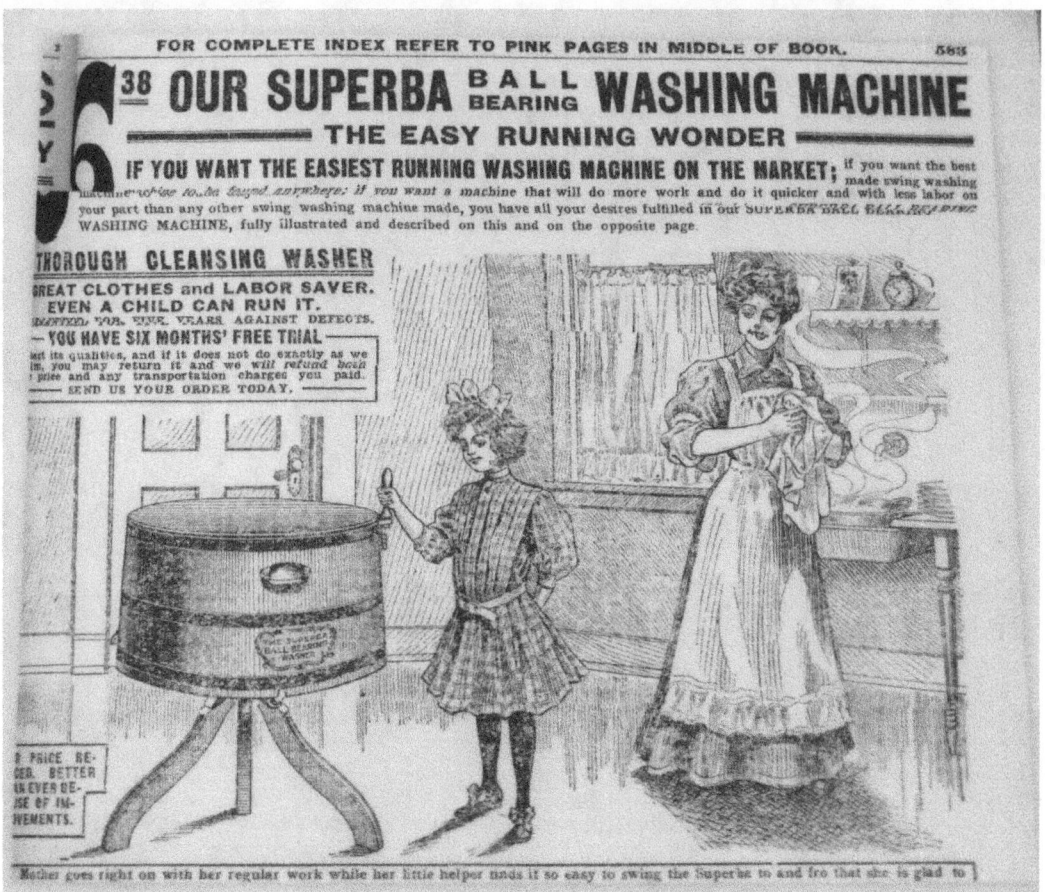

Every Monday was washday, then came the ironing on Tuesday. Not all women were homemakers. The mining communities had their share of businesswomen.

By the century's turn, San Juan merchants were stocking more of a variety of goods, but thanks to modern advertising competition continued to grow. In rapidly declining camps such as Dunton, Animas Forks, Eureka, and Ophir, the number of merchants declined just as rapidly. This left customers with little or no choice but to turn to catalogs and outside competition.

Even neighboring Durango represented a threat. The largest town in the San Juans, it offered a wider range of stores and businesses than any of its mountain neighbors. Its prices were also cheaper, and with both the Denver & Rio Grande and the Rio Grande Southern running in and out of the town, merchants could ship quickly and easily. San Juaners could also travel there to do their own shopping amid the delights of the big city.

A chair and a spittoon might mean a moment of repose. There was a growing movement, however, to end the tobacco and drinking evils.

Small businesses were not helped by the appearance of chain stores that could also undercut their prices and profits. Some were local merchants who expanded into several towns, but national chains loomed just over the horizon. These were not easy days for San Juan merchants, and they only grew worse as mining inevitably started to decline.

Customers had concerns and complaints about quality, variety, pricing, and service, as well. Further, merchants were not above price gouging. Rev. George Darley noted with a chuckle that Lake City merchants claimed the high prices resulted from the freight being so costly. Even needles cost ten times what they did in the East, because "freight was so high." Darley continued, "In those days, it [the cost of freighting] often meant a great deal more than the first cost."

Ouray newspaper editor Dave Day complained about prices in May 1880. When the price of flour reached twenty dollars per hundred pounds, paraphrasing Thomas Paine, he declared, "These are the times that try men's stomachs." It had seemed bad enough a couple of years before when that essential item reached twelve dollars for a hundred pounds.

Without question the cost of living in the San Juans throughout the mining days ranged higher than that in the foothills and in Colorado's larger metropolitan areas. Even with the coming of the railroad and the growth of agriculture around the mountains' southern and northern fringes, food and other essentials of life remained more costly. As long as times remained flush, that was not a heavy burden, but once decline set in, it only made the situation worse.

Hats off to all the cooks who worked against often extreme challenges to prepare meals. As the old saying goes, "Kissing don't last; cooking do."

"School days, school days. Dear old golden rule days."
Not only were schools important for education, but they enhanced the community's image.

CHAPTER SIX

Youthful Days, School Days

*The period of childhood may be taken as extending from the age
of two or three years up to puberty. Now what is wanted at this stage
is not so much acquisition of muscular strength or skill as a solid
foundation of general health.*
—Henry Northrop, *Golden Manual,
or the Royal Road To Success* (1891)

*Some of the kids were interested in learning—
some were very bright, some very dumb.*
—Annie Laurie Paddock,
Creede teacher, 1914–15

*I spent many hours making bouquets of clover blossoms and I
enjoyed walking on top of the board fence around our house.
In school, we played hopscotch, both round and straight.*
—Martha Gibbs, Telluride,
describing turn-of-the-century childhood

Children in mining camps and towns had a better possibility of being a Tom Sawyer or Huck Finn than a Becky Thatcher or some of the good boys Tom and Huck despised. They saw and heard things their Victorian nonmining community cousins often did not, or, if they did, were shocked by such carryings-on. It might have been preferable by Victorian standards, and was possible if extremely vigilant parents watched their children, not to have them come in contact with all aspects of life in and around their childhood mining days.

Generally, children matured rapidly in this atmosphere, and childhood vanished by the time young people reached their teens. By then, girls were becoming eligible to be courted, wooed, and married, and young boys might be working. Yet the world of the mining town must have been fascinating for young adventurers emulating a Tom Sawyer or Huck Finn because many adventures beckoned, and there was much to see, do, and learn.

This fascination was countered by the warning Alexander Gow clearly expressed in his 1873 book, *Good Morals and Gentle Manners for Schools and Families*. "There are persons with whom we must not associate; there are places where we dare not go; and there are things we should not see or know, if we would preserve our purity and self-respect." A definite tug-of-war existed between Victorian idealism and practical reality in the mining West.

This was not the only problem that caused parents to worry. Childhood diseases were a constant concern for every parent, particularly in smaller communities where a doctor might not be available. These topics are discussed in the next chapter.

Meanwhile, the school worked to educate frequently unappreciative youngsters. While not always practical in a transitory mining camp, the school building served as an important civic symbol. If a camp did not have one, it wanted one, as an 1880 letter from isolated, tiny Sherman bemoaned. When Capital City completed its school, the Lake City paper congratulated its neighbor on a "very neat structure, a substantial building warm, and well lighted." Most school buildings were wood, and all the students were packed into a room or two.

The pride of some of the larger towns was their brick or stone edifice, with individual rooms for a grade or two. By the century's turn, some towns contained high schools, but most students probably finished eight grades at best. For many parents no more education seemed necessary; it was a luxury to go on beyond grade school. Regardless, brick or stone, a school told visitors that the community was up-to-date educationally.

Teachers in mining communities faced almost impossible standards. Gow felt people expected, among other attributes, that teachers, male and female, would be "learned, wise, prudent, gentle, long suffering, impartial, energetic, polite and diligent." Then he added, "Such expectations are never realized."

School boards did expect teachers to toe the mark and be exemplary citizens. An 1873 board that, for instance, proclaimed that "any teacher who smokes, uses liquor in any form, frequents pool or public halls, or gets a shave in a barber shop will give good reason to suspect his worth, intention, integrity and honesty." Women were not immune to such generally stringent guidelines. "Women teachers who marry or engage in unseemly conduct will be dismissed."

Report cards pleased many and disappointed a few. Mabel certainly did well in March 1907, as excellent and good marks dotted her card.

It was no better in the second decade of the twentieth century. Women were not "to smoke," and under no "circumstances [should a woman] dye [her] hair." A woman teacher was not to "loiter downtown in ice cream stores" or ride in a carriage or automobile "with any man unless he is [her] father or brother." If a woman became in a "motherly way," in a time when sex was not a subject about which boys and girls needed to know, she would face dismissal or not be rehired.

Annie Laurie Paddock discussed what was expected of teachers in Creede in 1914–15. Women teachers had to "follow certain standards," which included teaching Sunday school. She joined the Literary Society that year and recalled it used as a text a history book whose author declared, "There would never be any more wars." Bridge was an acceptable game to play. Paddock enjoyed playing and did so despite having to walk home at night

Mechanical banks may not have improved saving methods, but they certainly entertained children and adults alike.

through Creede's unlighted, rough streets. Annie Laurie found out about life in a declining mining town when she was given a choice between living in a place with electric lights or one with a bathtub. She selected the latter but chuckled when she said the water was usually cold.

Most teachers in those years were women with a male superintendent or principal shepherding the system. They received the title professor, but the women did not. It was not equal pay for equal work; men were paid more. Depending on the time and place, the difference might range from twenty-five to seventy-five dollars a month.

Paddock, who taught third and fourth grades, described the experience. Not all the students had the same books; they had books someone gave them, and "they were old" and varied in reading level. Some of the "kids were interested in learning—some were very bright, some very dumb." She divided the two grades for spelling, arithmetic, and other subjects, but they "did some things together." She remembered being paid seventy-five dollars a month, and she "boarded on her own."

Local women provided the main support for the school. Paddock did not think the city government did "much about education," and the PTA existed to generate support.

Almost every schoolroom had a picture of the Father of the Country, along with one of Abraham Lincoln.

Girls' athletics were finally being allowed, and she coached the high school volleyball team. When talking to the author, she commented that the athletic field where the boys played was "full of stones and cobblestone and dust" and was a "mess."

Lake City's school board adopted rules for students as well as teachers. The school opened in 1875, and by 1877 the board thought it needed to give the young scholars some direction. Therefore, three unexcused tardy "marks equal a half day's absence," and five half-day absences "shall expel the pupil from school."

In the classroom, the *McGuffey Reader* probably remained the most popular book, with its levels corresponding to eight grades. As the decades passed, though, new educational theories and books challenged its dominance. Teaching, meanwhile, stressed the old favorites—reading, writing, and arithmetic.

The schoolhouse might be built, but progressive communities went further and planned to organize a school district. Lake City first petitioned the county superintendent of schools, who approved, and an election was held in November 1876. It passed, but some of the elected members failed qualify according to state requirements, so the process had to start over

She looks as perky today as she must have a hundred or so years ago, when she gladdened some young lady's heart.

again. Rico had the same problem, but eventually it all came together in both communities.

By the 1890s, some of the school districts had a fairly large number of students. Ouray's had 279 enrolled from primary through high school. For those who went beyond eighth grade, graduation day meant a special occasion. Telluride's 1907 graduates, all four of them, gave orations. The fascinating topics included the "problem of immigration," "little laborers of great cities," "the rise of Colorado," and "socialism: a menace to our government." That same year twelve graduated from Telluride's eighth grade, reflecting success in one of the largest school systems in the San Juans.

Teachers' salaries proved a vexing problem among the cost-conscious voters. By March 1877, with Lake City's teachers' salaries languishing nearly three months in arrears and the district $500 in debt, a ball held to raise funds netted only $36.50. That left the board with no other choice but to levy a tax. The editor of the *Silver World* weighed in: "Public schools should be maintained no matter whatever else may be neglected. We know the people of Lake City will not allow" the public schools to be suspended.

Mary Mott, who went to school in Lake City, remembered her teachers as running the gamut from mediocre and transient to "wonderful educators." A newspaper complained on one occasion that the students were not learning to spell. While there were poor teachers, others received praise from grateful parents. Ouray teachers in 1892 were hailed—"never in the history of the Ouray school" had they offered such satisfaction "to pupils and parents."

Telluride's teachers must have been doing well too, because the *Examiner* (September 11, 1909) noted that nine former pupils were in college, one of whom attended medical school. Other San Juan communities could not make such a claim at a time when going beyond high school was so rare.

At school, children playing could be an accident waiting to happen. On a cold November day in 1883, little Walter Kellogg had the misfortune of colliding with a "fox or a goose" ("he did not know which") and broke his collarbone. Fortunately, Lake City had a doctor.

Newspaper editors enjoyed printing school news. Lists of pupils with perfect attendance; those who were not tardy; and students' standing in such subjects as spelling, arithmetic, geography, and grammar made parents proud. The Hucks and Toms among their classmates probably exhibited less enthusiasm.

A few out-of-step people thought too much education could be harmful. Author Henry Northrop, for one, wrote in his *Golden Manual, or the Royal Road To Success* (1891) that "children at the present day are too highly educated—their brains are over-taxed, and thus weakened." His recommendation: "let him spend as much time as possible out of doors; let him spend the greater part of every day in the open air."

Northrop need not have worried about boys in mining communities not spending time out of doors. The camps and towns offered a virtual playground, with an almost infinite number of things to do and explore, although always with a certain degree of danger present.

Playing around open mine shafts, or in abandoned tunnels, led to more than one tragedy, as did playing with explosives. Fireworks could also be dangerous, and abandoned buildings, no matter how enticing, contained hazards.

Boys were generally given much more free rein to explore than girls, who found themselves more constrained by Victorian expectations. A proper young lady would not dream of doing certain things, but a few tomboys broke that mold—much to the shock of other parents.

A boy and his dog were a familiar sight. Girls could have pets—a cat, a bird, something less rambunctious. In fact, dogs overran the mining world, much to the dismay of residents and city councils. As the *Silver World* complained on September 18, 1875, a "goodly number of dogs" seemed to have a "peculiar mission in life" to "disturb our slumbers." Rico had a similar problem, but the editor of the *Dolores News* approached it differently. By "careful and accurate account," there had been "13,697 dog fights on the streets" in the past week. One well-known pooch "Carlo, lost the championship on Thursday night."

If dogs created noise, so did the "burro glee club." One reporter observed, "The bray of the gentle burro makes nights hideous." Cats caused fewer problems and along with some dogs were looked upon as a definite necessity to help keep down the mouse, rat, and other small-rodent populations.

Like people in other eras, San Juaners became quite attached to their pets. When Old Tom, the cat in Telluride's Kracaw Grocery, died the *Examiner* was moved to say that Tom greeted everyone, "whether kept in the red or black ink on the books. He will be missed."

There was nothing like a circus to stir excitement among the children and kids at heart. With the coming of the railroad, circuses could easily travel into the San Juans. McMahon's & Farini's Circus visited Ouray in August 1889, to the applause of editor Dave Day. A pair of burros "did the best act," he thought, but the "performances of a dozen horses" were "the

The 1897 Sears, Roebuck and Co. Catalogue offered wagons from $1.15 to $6.75. This style ranged from $1.15 to $1.65.

finest he had ever seen." Day, who obviously liked the circus, also raved about the trapeze, horizontal bar, and contortion acts.

Holidays were special times for children. Even the poorest families tried to have some toy, book, or clothes in their children's Christmas stockings. Wealthier families indulged their children with more gifts and perhaps a tree with candles, sometimes a dangerous combination.

The variety of toys available, locally or from a Sears or Wards catalog, might surprise modern parents. Dolls, books, bicycles, baseballs, candy, and wagons were offered, along with toy safes, carpet sweepers, brooms, sleds, and trunks, as well as hobbyhorses and a large variety of games. Clothes arrived with Santa, something girls probably appreciated more than boys.

With plenty of cold weather and snow, old and young alike could enjoy coasting on sleds, ice skating, and snowshoeing, as skiing was called. Mining engineer Eben Olcott wrote his sister about a coasting evening at

A variety of toys, some dating from the World War I era, are pictured. Sears offered gloves from 85¢ to $3.60 for a "baseman Mitt made of specially tanned calfskin."

Lake City. "Everything was in prime condition and the full moon made it easily as bright as day.... We fairly fly," going down a steep hill, he wrote. "It makes it very exciting."

Some enjoyed throwing snowballs, but not their victims, particularly adult males who did not appreciate being the target when some rascal tried to knock off their hats. Firecrackers also caused problems, particularly when thrown around horses or other animals. They also inflicted injuries on occasion on those who played with them.

Children in the mining world came face-to-face with a series of vices, from drinking to smoking to drugs. There is no doubt they were tempted and might have sampled these forbidden pleasures. Telluride teachers, for example, were "up in arms" in 1909 over "the contagion of cigarette smoking by many grade school pupils."

The *San Miguel Examiner* (November 6, 1909) thought the problem was serious enough to write an editorial aimed at both the young and the old. "The cigarette smoking habit could and should be overcome not only in children but men as well." The editor continued, "The school is a splendid place to begin to [teach] the discipline for that habit which has never done anyone any good and will grow upon a person with age."

It probably did little good, as youngsters watched their elders chew and spit tobacco and enjoy their pipes, cigars, and cigarettes. The same might have been true with drinking, although churches and the Woman's Christian Temperance Union fought valiantly to show the evils of demon rum and its nefarious friends on those who indulged in such evil temptations. Youngsters signed pledges and joined groups, but the lasting effects are unknown.

City councils tried to regulate temptations and dangers for the sake of local children. The red-light district obviously was off limits, with fines and a curfew after which children could not be downtown and fines for parents who allowed their offspring to wander there. Rico, because of the great danger of coasting, prohibited the practice on certain streets and "the depot hill." Throwing firecrackers under or around animals was prohibited in all towns, with fines and even the possibility of days in jail. Throwing stones, snowballs, or any "missiles" against any person could lead to a two-dollar or more fine in Silverton, although offenders more likely only got a lecture.

Contagious diseases also caught the councils' attention, as discussed in the next chapter. Citizens and councils prohibited riding bicycles on sidewalks. The rather vaguely defined "disturbing the public peace" might get a young man a lecture or a trip to the woodshed with his father. In the Victorian world, a young lady would likely never have engaged in such behavior.

Children in the mining camps matured rapidly, although they probably enjoyed their childhood. As Mark Twain wrote in the 1876 preface to his classic *Tom Sawyer*, "Part of my plan has been to pleasantly remind adults of what they were themselves, and of how they felt and thought and talked, and what queer enterprises they sometimes engaged in." Twain understood that kids will be kids, no matter what the era, and adults will be adults.

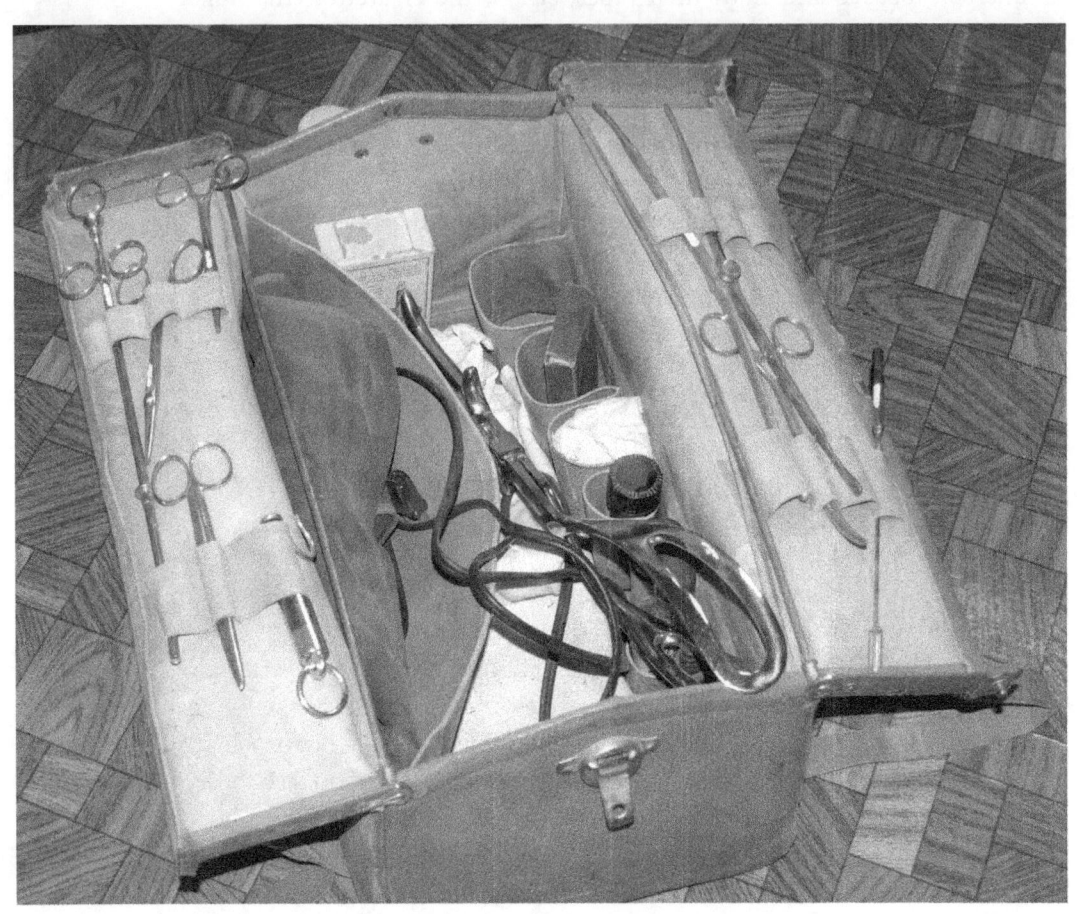

Patients hoped that a doctor's bag, with its pills, potions, and tools, contained the answer for their complaints.

CHAPTER SEVEN

Dentists, Doctors, Disease, and Death

*Probably as many lives have been saved by
good nursing as by good doctors.*
—Henry Northrop, *Golden Manual,
or the Royal Road To Success* (1891)

*Dr. E. W. Sheriff, one of Colorado's most skillful dentists
will be in Ouray this coming week. Parties desiring work in
the dental line will do well to call on him.*
—*Solid Muldoon*, October 3, 1879

Being ill in a San Juan mining town or camp was a frightening experience. Typically, no doctors practiced in the camps, and many people had serious questions about nineteenth-century physicians, as the *Golden Manual* stated. Further, the high elevation produced complications people at sea level seldom faced, although some folk proclaimed that living in the mountains actually produced healthier lungs and bodies.

Some problems stemmed from the fact that these years represented a transitional period in medicine. Diagnosis was improving, medicines were a blend of old and new, doctors were gaining new insights into bacteriology, and awareness of the importance of sanitary conditions in operations and general practice was slowly leading the way to modern medicine. Fortunately, anesthesia for surgery provided a major breakthrough from the pain-filled earlier days.

Medical schools with degree programs had been founded, and graduates were appearing. Most graduates, however, went to the cities where incomes were higher and living was easier. Almost anyone could put doctor in front of his or her name and start a practice, though. Some had trained under another doctor for a short period (weeks) or longer (a year), others read books, and some had done nothing but become fascinated by the profession. There were few medical licensing laws or regulations at the time, so it remained "patient beware."

Rural America was quite often short of doctors, whether on the Illinois prairies or in the San Juan Mountains. Many Americans did not trust doctors or hospitals, which were places you went to die. What they did rely on were home remedies or the popular patent medicines. To continue the *Golden Manual*'s recommendations: "Medical skill cannot always save a life, but it has a far better prospect of doing it when accompanied by proper care for the sick."

For those who were ill, the book recommended "well prepared and cooked meals," a clean sickroom, plenty of sleep, and "medication to the smallest possible quantity." Depending on the illness, but certainly during recovery, suggestions included "big doses of sunshine" and keeping the patient calm at all times.

Home remedies passed down from generation to generation had arrived in the San Juans with the pioneers. It seemed that for some people, strong-smelling and vile-tasting concoctions proved the best.

Some of the remedies worked, some probably provided no help or harm, and some must have hurt the patient more than they helped. Mustard plasters for a variety of aliments, whiskey or brandy or both for snakebites, cold water for sunstroke and other emergencies, unsalted butter or egg yoke on burns, and coffee grounds or a handful of flour bound on a cut to stop bleeding are a few examples. Foul-smelling asafetida bags, hung around children's necks, supposedly produced medical miracles but left a distinguishing odor. For an earache, one remedy suggested soaking cotton in a mixture of black pepper and "sweet oil" and inserting it into the ear.

Gargling turpentine or kerosene mixed with sugar or honey for sore throats does not seem an enjoyable procedure, but rubbing horse liniment on aches and pains might have helped. Should one be struck by lightning and happen to survive, whiskey and aromatic "spirits of ammonia" were recommended. Almost everyone suffered with an aching tooth at one time or another and for that, a piece of cotton saturated with ammonia and "placed on the defective tooth is excellent."

Some remedies could be fatal. Turpentine to treat dysentery was like throwing gasoline on a fire. Quinine was considered a miracle medicine and opium the queen of medicines. Both could be easily purchased and applied in any dose imaginable to cure, it was claimed, a variety of illnesses. Sadly, the latter and its derivatives also caused addiction.

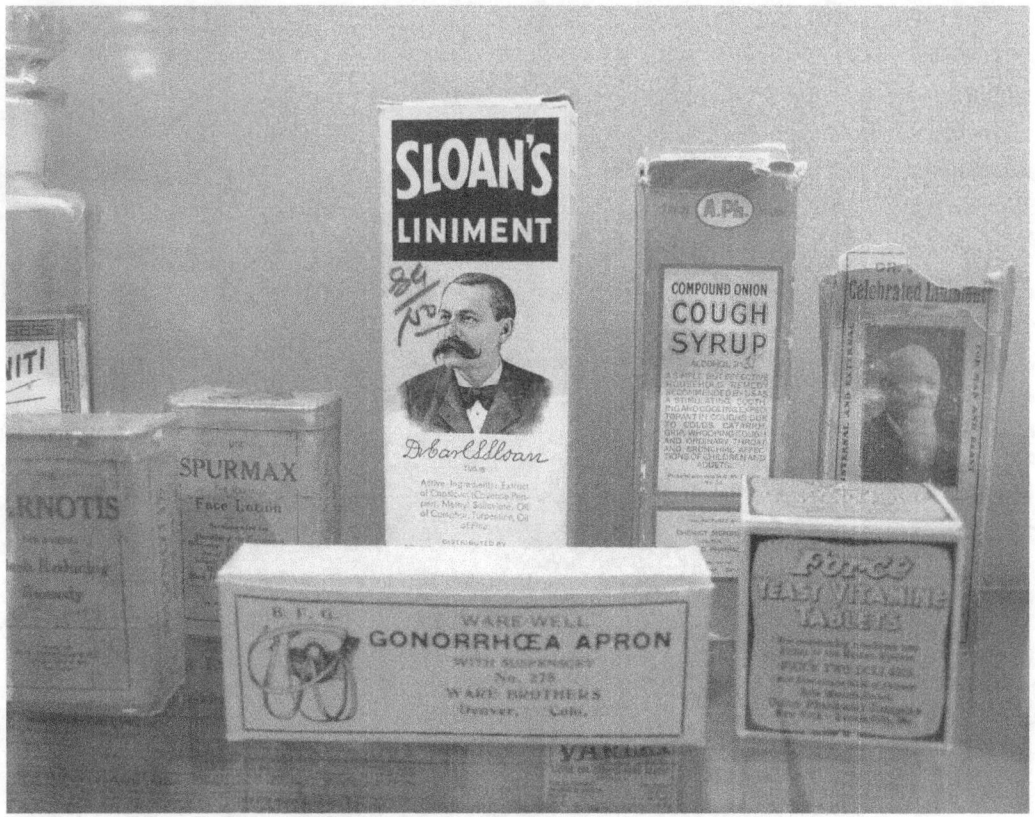

Sears catalog offered eleven pages of patent medicines. The company promised that "if you cannot find what you want write to us."

One did not need doctors because "better wonders" could be found on local store shelves or ordered through the mail. Patent medicines offered miraculous cures for almost all known aliments. Mining-camp folks of all ages were promised relief for whatever ailed them. According to advertisements, there was no cause for them to suffer when a cure lay at their fingertips. In this time before enactment of the Pure Food and Drug Act (1906), patent medicine advertisements promised marvels in pills or bottles, miracles that the local doctors could not accomplish.

To make matters worse, advertising ran unchecked and was inviting. No agency examined the claims to see if they were truthful. It was truly an era of "buyer beware."

A sampling of remedies available provides an indication of the needs and hopes of apprehensive clients and illuminates what people suffered from, or thought they did. Dr. Chaise's Nerve and Brain Pills were "positively guaranteed to cure any Disease for which they are intended,"

A cure for almost any ailment known to men and women came with one of Hahnemann Health Restoratives.

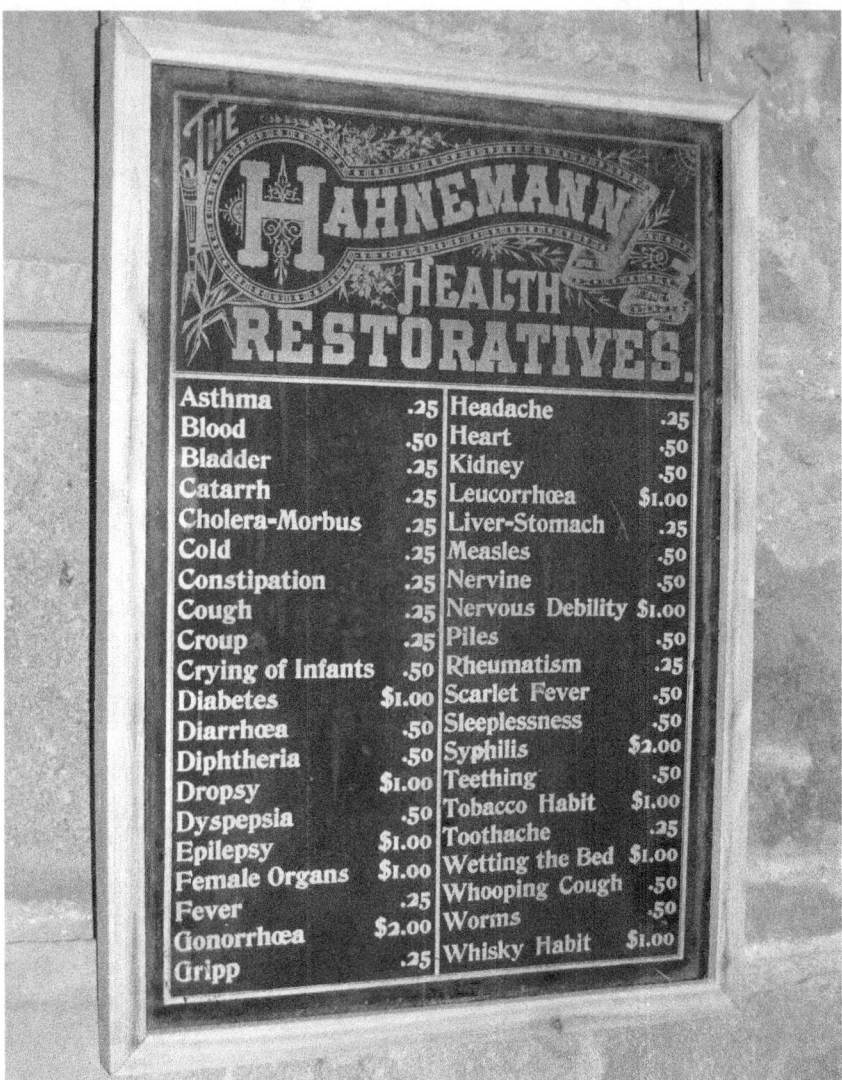

including "impotence, spermatoranga [*sic*], overwork, sexual excesses, etc." Dr. Baker's Blood Builder proclaimed it to be "nature's most wonderful remedy for destroying poisons in the blood and building up a pure healthy blood, no matter how diseased the system is." If the doctor label was not present to give authenticity, some exotic cure from China or Turkey or a host of testimonials from the "cured" would suffice.

The "Positive Rheumatic" cure made "the lame well," and Dr. Rose's Obesity Powders solved the problem of "too much fat," which was "a great annoyance to those afflicted." Suffering from "indigestion and dyspepsia"? Dr. Wilden's "quick cure" was "the stomach remedy." "Reliable Worm Syrup" solved that problem with children. "Pasteur's Microbe Killer"

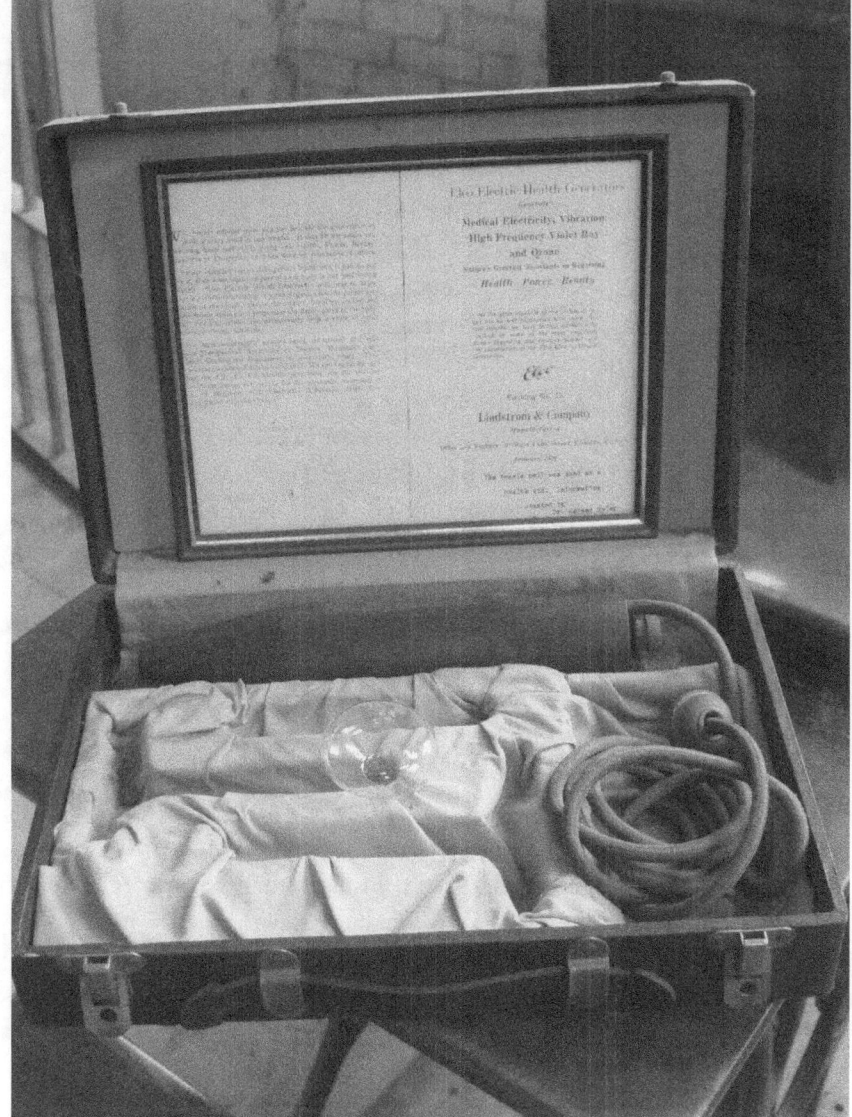

If Hahnemann's did not work, try electricity. As one advertisement boldly stated, "The Electro Medical Battery as a curative is becoming more appreciated from day to day."

would "prevent LaGrippe [flu], catarrh, consumption, malaria, blood poison, rheumatism and all diseases of the blood." Dr. Rose's Kidney and Liver Cure came "guaranteed the greatest kidney and liver remedy ever known." German Herb Tea offered "a positive cure for constipation," and the Mexican Headache Cure "guaranteed splitting headaches cured."

Lofotem Cod Liver Oil promoted itself as the "greatest remedy ever discovered" for a host of problems—coughs, skin diseases, diseases of children, general debility, colds, phthisis, and rheumatism. It was a medicine cabinet all in one bottle for only sixty cents. For another fifty cents, the German Liquor Cure could cure the "liquor habit" so that "every man can be permanently cured of the habit or desire for intoxicating drink of any kind." Women, of course, would not dream of indulging in the devil in a bottle. Almost all these wonders cost less than a dollar.

Laudanum could be purchased in stores or from Sears and was hailed as "always useful [for] both children and adults." A host of medicines surfaced just for women, including Lydia Pinkham's Vegetable Compound (probably the most famous patent medicine of the time), Dr. Worden's Female Pills for Weak Women, and Brown's Vegetable Cure, "the weak women's friend." Together, they managed to solve all female problems, or so it seemed.

No doctor could match these cures and promises of relief from suffering. They faced an uphill struggle to convince patients that the advertisements claimed more than the supposed medicines could produce. Few mentioned that many of the remedies contained opium or some other addictive drug. A few advertised that they prescribed "no opium or other injurious substances" because by the 1890s, physicians and others were becoming aware of the addictive properties of some of the proclaimed miracle drugs.

Opium was not the only problem. Many of the remedies contained alcohol, which made the sufferer feel better, at least temporarily. Again, the purchaser did not know this, and the WCTU (Woman's Christian Temperance Union) actually endorsed Pinkham's Compound—only to learn later the principal ingredient was alcohol (some wags said that "there was a baby in every bottle").

Born in 1819, Lydia Pinkham concocted her compound for women and created a company that manufactured the most widely advertised product of its time. She carried her advertising one step further, promising that anyone who wrote her with a problem would receive an answer. Long after her death, a roomful of clerks still faithfully answered queries in her name.

No mining-community doctor could compete with such wonders. Further, who can say that the patent medicines did not do some good? If they did not help physically they may have done so mentally, as simply taking them perhaps offered some relief in the sufferer's mind.

Traveling dentists and doctors appeared in the San Juans in the 1870s, and some settled in the towns. They often traveled around as much as their patients, however. Dr. J. Blake practiced in Silverton, Rosita, Lake City, and Animas City before finally settling in Rico. Ouray was fortunate to have two physicians by 1878, while smaller camps around Ouray and Silverton had none.

Sickness took its toll, and many reasons were advanced as to why. Rico's *Dolores News* (November 22, 1879) blamed an outbreak of colds, sore throats, and influenza on "too much comfort" and "tightly roofed, very comfortable habitations." The editor thought hot air was causing the trouble and recommended making an opening in the roof to "correct the evil."

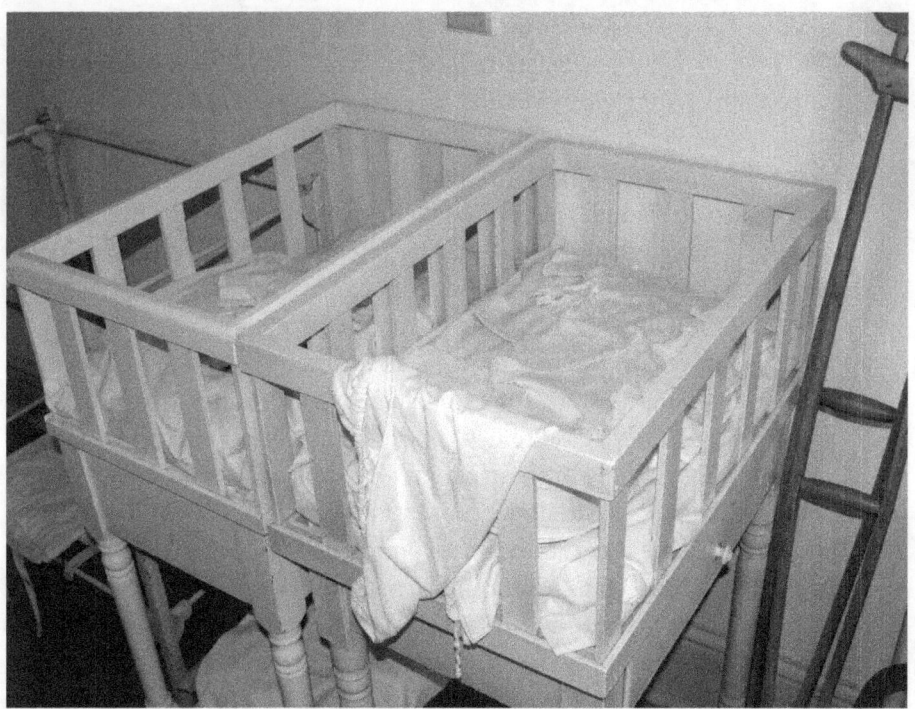

Most babies were born at home, but San Juan hospitals had their baby cribs ready and waiting. When Telluride got a hospital, Silverton was beside itself with jealousy.

The altitude was also blamed, particularly for pneumonia deaths. The *News* reported in April 1880 that "doctors have but little power to successfully battle the disease in this high altitude." The editor recommended "caution" and avoiding "exposure," for "18 to 36 hours is sufficient to lay the heartiest of us in the grave." Smallpox, despite inoculation being a well-known preventive, still stalked San Juaners.

Doctors did much of their work, both diagnosis and treatment, through house calls. That way patients stayed in a familiar environment and did not come in contact with anyone else who was sick. It was not unknown for doctors to perform operations on kitchen tables, and most babies were born at home.

Despite some people's reluctance to go to hospitals, by the 1880s Ouray, Telluride, Lake City, and Durango all had them, or what was professed to be one. The Sisters of Mercy started two of those institutions. In the years ahead Silverton acquired a hospital, and Telluride gained a second one thanks to the miners' union. Creede opened a hospital amazingly early

> *Mark Twain believed "the average man dreaded tooth-pulling more than amputation." The introduction of nitrous oxide (laughing gas) improved the visit.*

in its history, in January 1892; the facility claimed to have a "full corps of surgeons." With a practical bent, it was noted that "until a crop of patients is harvested, it will be used for a hotel."

Meanwhile, Ouray and Durango advertised their hot springs as miracle cure-alls. Indeed, in the nineteenth century many considered them such for ailments ranging from sour stomachs to aches and pains. Some promoters asserted they could renew youth and do almost anything else they could dream up. Some of these claims were as bad as those noted earlier for patent medicines. No doubt, however, they might at least temporarily have offered some relief from those common complaints of joint pains or rheumatism. The water's heat could be soothing by itself.

In December 1898 the *Ouray Herald* launched a crusade against "quacks and charlatans" advertising in newspapers, purporting that physicians "ought not to stoop to this method." The editor was unsuccessful, in part because such advertising was an important financial windfall for hard-pressed papers.

San Juaners' mouths likely approached dental disasters. Few practiced even the simplest dental hygiene—brushing their teeth. When they did see a dentist, if one were available, it was typically to have a tooth extracted or address some other crisis. If no dentist could be found, someone else, perhaps a blacksmith, yanked out the offending tooth.

Dentists were found in some of the larger towns but less frequently in camps. As Dr. Sheriff's notice, quoted earlier, pointed out, traveling dentists filled cavities, pulled teeth, and provided other needed dental services. A visit to the dentist could be a harrowing experience. Dental equipment and techniques had a long way to go before they minimized inconveniences and solved problems. Some folks preferred a shot of whiskey to calm an aching tooth, which in the long run solved nothing.

Waltus Watkins, a dentist at Amethyst near Creede, wrote about his experiences in 1899. Many people wanted "gold crowns and bridges, & some want gold crowns on front teeth for display. They ought not to be put there of course, but I am not here to regain lost health. I crown them when they want it done." He worked in an office with two "incandescent lights in front of my dental chair; each light is 32 candle. I do considerable night work for miners." Realizing that Creede had slipped past its prime, Watkins was gone within a year and relocated in Craig, Colorado.

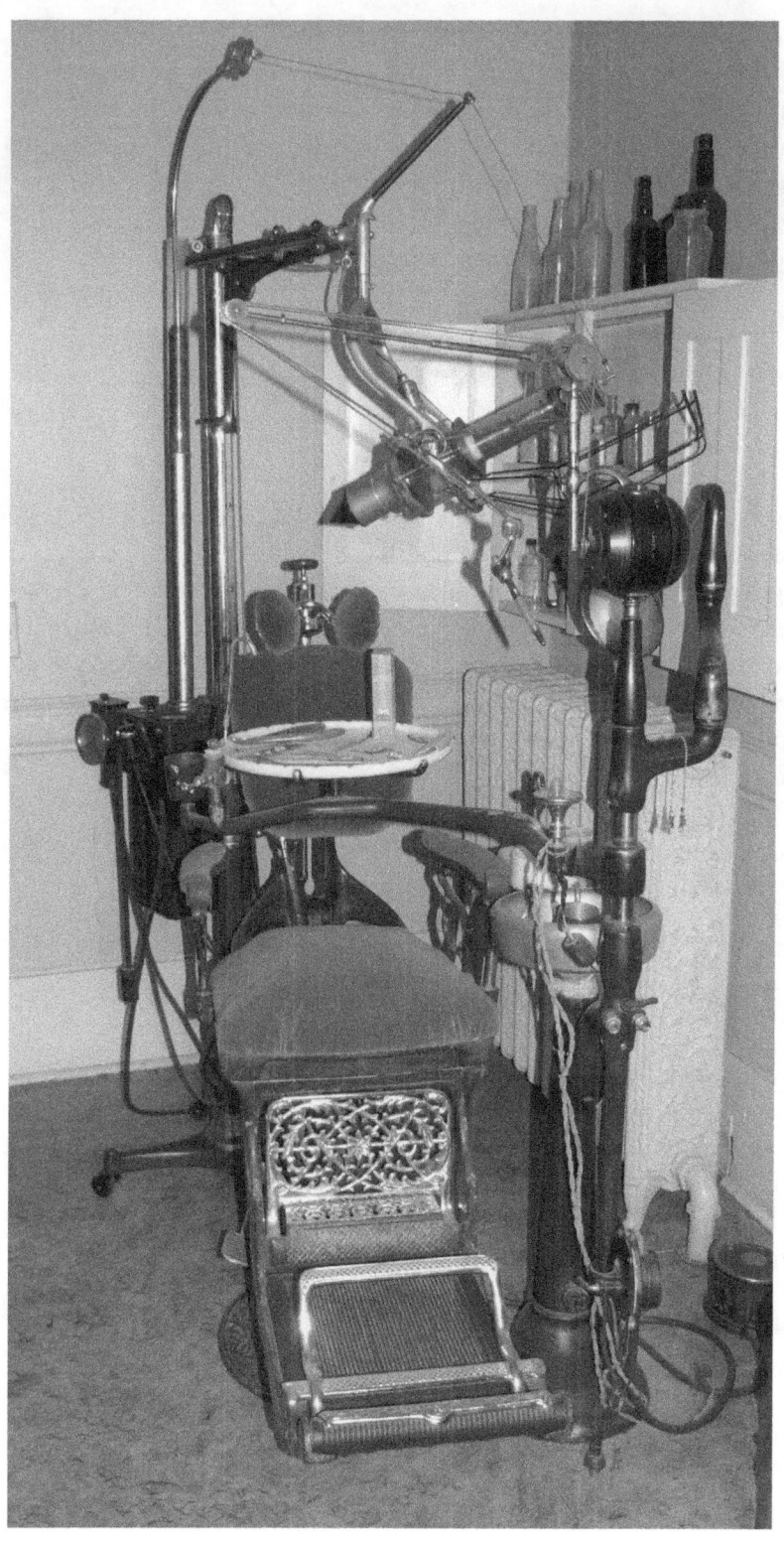

For patients, pain was always the major worry, as the drills were slow and pump-operated by the dentist's foot. Only during the late nineteenth century did the miracle of painless dentistry arrive with the advent of nitrous oxide, or laughing gas. At high altitudes, however, it led to a variety of problems, including causing the patients to talk, laugh, and do hilarious things—much to the amusement of onlookers who happened to be in the vicinity. Many patients agreed with this sonnet about a dental visit.

> *Who seats me in his easy chair,*
> *And hurts me more than I can bear,*
> *And pulls my tooth and doesn't care?*
> *My Dentist.*

Death loomed everywhere in the San Juans. Frank Hough, in his Lake City diaries of 1900 and 1901, recorded a number of deaths of babies and youngsters. An old saying went that a family should have four children—one for mom, one for dad, one to increase the population, and the fourth to die. Such childhood diseases as chicken pox, mumps, and diphtheria could be fatal or lead to fatal consequences. Pregnancy and birth could be deadly for both the mother and the newborn. As Thomas Hornsby Ferril wrote in his poem "Magenta":

> *Each woman had seven children of whom two*
> *Were living, and the two would go to church.*
> *. . .*
> *Maybe you never saw a miner dig*
> *A grave for a woman he brought across the plains.*

With no reliable birth-control methods available (those "rubber things," as Victorians contemptuously called them, certainly were not) except abstinence, babies arrived frequently. For both mother and child in the San Juan region, the isolation, high elevation, poor sanitation and water, and often poor housing, along with a lack of skilled doctors trained in women's and children's medical needs, meant trouble.

It did not help that the mining communities were unhealthy places. Clean water remained a continual problem, often growing worse as mine dumps and outhouses polluted streams. Filth was found everywhere in the towns—from outhouses to garbage to horse and mule chips to litter carelessly thrown about. Animal carcasses were sometimes left to rot, and the air became polluted in a variety of ways. Bathtubs remained a rare commodity, and overall "cleanliness" did not quite rank "next to godliness." A

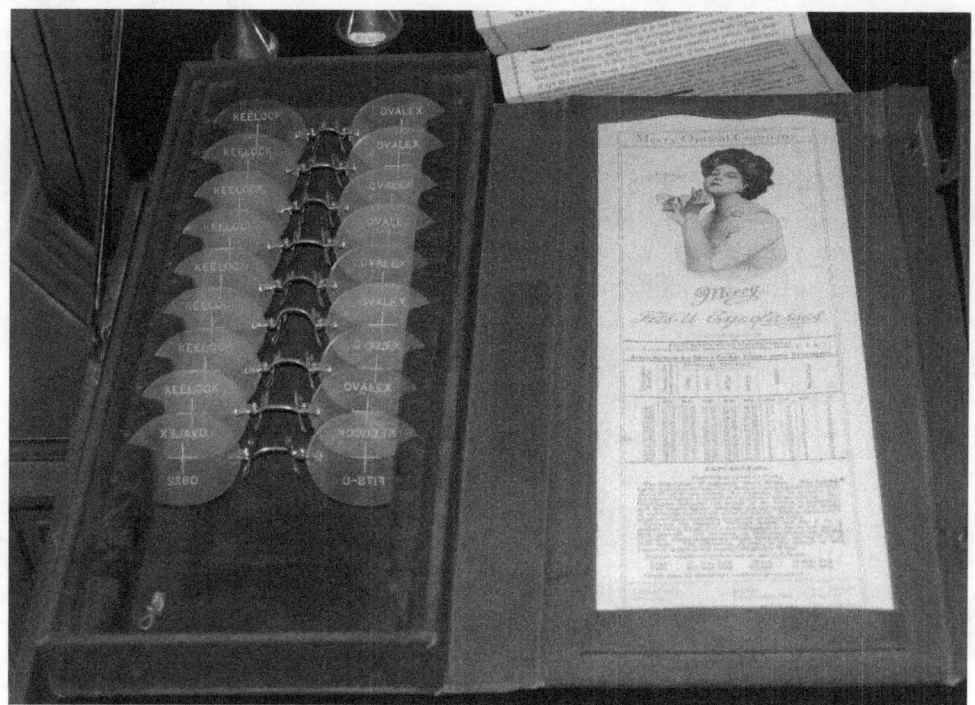

In order to be sure one could clearly see this beautiful young lady, a new set of glasses provided the perfect answer.

modern visitor would have been offended by the smells and general physical appearance of these camps and towns.

A host of diseases and disasters plagued San Juaners. Common cuts could become infected and result in blood poisoning, hydrophobia was incurable, high-altitude sun caused a variety of skin problems, and colds might turn into pneumonia. Syphilis could not be cured, though its symptoms could be masked, and any infection or illness might threaten one's health if not treated quickly and properly.

Nature, too, claimed victims, mostly in the winter. San Juaners died from avalanches, exposure, rockslides, slipping on ice, and drowning in high water. They needed to be especially careful while traveling, because the mountains were unforgiving in any season.

Mining accidents took a toll, and so did miners' consumption (pneumoconiosis), contracted from breathing the dust the power drills generated. One woman wrote the author about her early recollections in Creede of "strong men suffering a slow death" from "knock in the box," as it was often called. A person might not die from it, but a related lung or respiratory problem would nearly always prove fatal.

George Darley, while ministering in Lake City in the late 1870s, listed consumption, death from freezing and exposure, mining accidents, fever, and the all-encompassing "found dead" as the causes of death for the people he buried. In Ophir, murder, suicide, pneumonia, snowslides, mining accidents, exposure, drowning, heart failure, and old age claimed their victims.

A sampling of the causes of death for persons buried in Silverton's cemetery during the years under discussion disclosed that mining accidents, snowslides, pneumonia, and heart disease proved the most prevalent for adults. Despite the western legends of gunfights, very few persons died from gunshots, and most of those were suicides or accidents.

It is hard to understand how modern "modern medicine" really is. It was not until well into the years under discussion that germ theory and the discovery of the causes of specific diseases came about. Other breakthroughs came later. Modern medicine and modern dentistry arrived with and after World War II.

Still, the folks in the San Juan mining camps survived despite a high mortality rate and living conditions that were often not the best. The few dentists and doctors available were always on call and tried to do their best in a wide variety of situations. "Three cheers and a tiger" to them all.

Pioneer minister George Darley observed that ministers "tire and tire again" in bringing the word of God to the "washed and unwashed" in the mining communities. Shown here, Ouray's church.

CHAPTER EIGHT

"Shall We Gather at the River" or "Shall We Go Straight to Hell"

Shall we gather at the river
Where bright angel feet have trod
With its crystal tide forever
Flowing by the throne of God?
—"Shall We Gather at the River"

I am bound for the promised land,
I am bound for the promised land;
Oh, who will come and go with me?
I am bound for the promised land.
—"On Jordan's Stormy Banks I Stand"

During and after the bloody Civil War, many people felt the need to sing religious songs whether back in the East or in the San Juans. Gospel songs, more to popular liking, replaced the stricter hymnody of prewar days in many churches, particularly those most likely to migrate west with those frontier faiths, the pioneering Baptists and Methodists.

Americans, as the popular evangelist Billy Sunday proclaimed, needed "to get right with God." Presbyterian George Darley was one of the first and finest ministers to venture into the San Juans. While he had to "suffer and suffer again," in his words, to bring the gospel to these rugged mountains, he persevered. By the late 1890s, Darley looked back with pride at how the church had taken root in an era that some thought bordered on being another Sodom and Gomorrah.

For Darley, the challenge was softened by the environment of the majestic San Juans:

> History teaches that Christianity generally prospers in mountainous regions. The length, height, breadth and grandeur of our great mountain ranges are in perfect harmony with the rugged, grand and sublime evangelical doctrines presented by ministers of the Gospel. The great peaks remind me of many of the precious truths presented in the Bible.

It took a certain type of minister or layman to reach out to the pioneering westerners. No doctrinaire, dogmatic individual would achieve great success. A person instead had to be willing to go where the people lived and worked, accept them as they were, and preach wherever he could and whenever he could. That characterized the type of individual needed. He had to be willing, for instance, to go to "hell's acre" to reach the women and men who resided and toiled there.

Darley recounted this in his story of conducting Magg Hartman's funeral. One of the denizens of the acre, he consented, "because I believed it my duty to go wherever I was asked for the purpose of conducting funeral services. As the 'girls' came in from the 'dance-halls' I took each one by the hand and spoke a kind word." They expected him to "wade" into them and their life, but he did not, and they relaxed—"soon tears began filling their eyes." An experienced minister, he understood reality. "Soon every head was bowed, and had I not witnessed such before, I might have believed every one would leave the paths of sin and seek a better life."

Young priest James Gibbon found the same experiences when he went to serve the Roman Catholic churches in Silverton and Ouray, as well as Roman Catholics throughout the mountains. It was August 1888 when he arrived at his first appointment. Looking back on those days, he confessed to his inexperience without specifying details:

> The success of my first year in the San Juan would have been greater, had the conditions of the parish been more favorable to harmonious action. Misunderstandings will arise in the best regulated societies, and to realize to fruitful purpose the divine constitution of the church, which distinguishes the teaching from the hearing element, needs a good will more than scholastic acquirements.

Gibbons learned, grew in his calling, and served faithfully for years.

"I lift up my eyes to the hills—from where will my help come? My help comes from the Lord," affirmed the writer of Psalm 121, who could have been referring to Silverton.

George Darley arrived in the San Juans in the 1870s. He organized and then helped build, with his brother Alex, the first church in the San Juans at Lake City in 1876. He gave young ministers the following advice: "On the frontier, among the highest ranges of the Rockies, you can find work that will try all the love for Christ that is in your heart. . . . Never fear that you will not have opportunities of showing whether or not your religion is the kind that will bear testing."

It might have seemed easy to get the Lake City church established, but it was not. Previously, local folks had tried to organize a Sunday school, often the first step toward a church. They ended feuding, and the entire project failed. The Darley brothers proved to have the talent, determination, and ability to bring everything together, yet even they did not always succeed.

Tested these men of God would be. Sunday was not the Lord's Day, as many still thought of it—a day of church service, Sunday school, and an evening prayer meeting. Sunday closing laws were standard in many communities back across the wide Missouri but not in the West. Sunday in the San Juans, and elsewhere in the mining West, was a wide-open day. With many miners and others coming to town on their day off, businesses found it one of their best days in the week. Entertainment of various types flourished, and the camp or town seemed to many visitors to have retrogressed from the Sabbath they had known back east.

The theme of a mining community reflected the materialism of the industry, often in its most aggressive form. Ministers had to comprehend this to understand their congregations. The West, Darley thought, had a negative influence on men. "They used to remember the Sabbath Day and to keep it holy back East," an attitude they "break in two out West." He continued, noting that folks had once attended prayer meetings but no longer did, nor did they volunteer to teach Sunday school as they once might have.

Darley pointed out disgustedly, "They are the kind of Christians who joined some church 'back East,' but who never joined the Lord Jesus Christ." He was certain they believed "the lord did not see them after they crossed the Missouri River" and did not want ministers to "bother them until they recross that muddy stream on the way back." Darley could only say, "They are the kind of Christians who are walking hand in hand with the devil, going straight to hell." Despite that feeling, the hard-working Darley never gave up trying to save one and all, regardless of whether they were in his flock.

There might be "bold infidelity" in the mountain towns, he concluded. Not one to run from a challenge, he also believed there "are many more faithful followers of Christ than some men like to have us believe."

There were other challenges as well, fortunately more humorous than serious. Darley remembered one time when a terrier dog and a tomcat "visited church" and walked into the choir corner. They were "soon at it and such a racket had never before been heard in the choir corner." A deer walked into church once during a service, and mules outside often joined in with their own music.

Ministers had to have a sense of humor. Darley recalled one time when everything was set up nicely for him in a saloon—too nicely as it turned out, because the "pulpit" proved to be a box filled with rotten onions. He preached the service anyway and turned the tables by "not seeming to notice that there was anything wrong."

He made another observation. "One peculiarity about drinking men in 'live' mining camps is their sense of honor when dealing with a minister; they insist on paying for their preaching and for funeral services." He concluded, "If Christians everywhere were as generous" it would "be highly appreciated."

Ministers had to be alert for all kinds of interruptions. Darley recalled a fight between a terrier dog and a tomcat. "Such a racket was never before heard in choir corner," certainly not in Sliverton.

Every minister who ventured into the San Juans had to keep these facts in mind if he hoped to succeed. If not, he might find himself with little or no congregation at first and without funds. Conversely, however, town boosters and boomers wanted a church—that is, a church building—regardless of whether many attended it for Sabbath services. A church, like the school building, provided a sign of permanency and civilization, a place people would want to visit, families to settle, and investors to invest in for the future. For many it was a civic improvement, not necessarily a place of worship on Sunday.

Despite all the pros and cons, ministers ventured into the San Juans almost on the heels of the miners. They were just as pioneering, even if they

The Reverend James Gibbons remembered that "the Silverton church workers were second to none in the state, and, strange to say, were nearly all women." Women generally played the church organ.

focused on goals of salvation and treasures stored in heaven, not gold and silver on earth.

Ministers and priests serving in the San Juans tired and tired again facing a variety of situations. One was simply traveling to and fro. The Reverend James Gibbons, who served several parishes in the late 1880s and early 1890s, as mentioned earlier, left this account of a stage trip to Silverton from Ouray to conduct a mass. He made the trip "without any more serious inconvenience," he recalled with a bit of tongue in cheek,

> than that of finding myself obliged to shovel snow, open the road and help drag out the horses from high drifts. Napoleon's trip across the Alps may be considered pleasant when compared with the fatigue and perils of a journey way up in the clouds during one of the fierce storms which sweep through the canyons. At times it is hard to tell which way the wind blows; it comes at once from all points and so thick is the fine sifted snow that you are almost blinded.

Once they reached their preaching destination, there was most likely no church building available except in the larger towns. They had to make do, and Darley was a master of doing just that—even in a saloon:

When entering camps where no religious services had been held I invariably went to the right place to find an audience; and in every case was courteously and kindly received and generally told: "Just wait, Brother Darley, until the games can be stopped, and we will give you a chance at the boys." It is not always an easy matter to stop the games; winners were usually willing, while losers were not. But so soon as the games closed the "roulette," "keno," "poker," and "faro" would give place for a time to the Gospel.

The men hung out at the saloons, and there was no better place to reach them, although conservative, eastern divines might have take umbrage at such a place for worship among a congregation of sinners with all those gambling temptations.

Despite the feeling of some people that "God had been left back on the Missouri," ministers and lay preachers had arrived in the Pike's Peak country in 1859 and reached the San Juans in 1874. To give the first sermon at Howardsville, at the mouth of Stony Pass, Benjamin Crary endured a whole week of hard travel from Del Norte—a distance, as the crow flies, of slightly less than a hundred miles, but the crow does not fly straight in the rugged San Juans.

John Wesley admonished his Methodist followers to "sing in time. Sing lustily and with a good courage. Sing modestly. Do not bawl."

The Darleys and others had similar problems getting in and out of the San Juans in the 1870s. The exception was Lake City, which set on the eastern fringe, not in the heart of the mountains. It was logical that the first church would appear there.

Lake City exhibited typical pride of gaining a church building, for to people of the day, the building was a positive sign of progress. The *Silver World* (October 28, 1876) proclaimed that "Lake City is always in the van in enterprise and public spirit and exhibiting unexampled growth and well founded prosperity in business." The editor continued, "Now it can make the claim few older towns and no other town in the San Juan country can make—having a well appointed and fine church edifice."

The article went on in great detail about the church, including the twelve-hundred-dollar cost, how the money had been raised, and that the fact it would be alternately used by the Presbyterian and Methodist denominations. Lake City was proud and wanted the world to know of its accomplishment.

Women had undertaken the task of raising money—about three hundred dollars—for the furnishings. With enthusiasm and determination, they had accomplished their goal. This role for women was not unusual; they, along with the minister or priest, provided the backbone of the church. Gibbons pointed this out thankfully, if somewhat amazed:

> I said that the church workers were, strange to say, nearly all women. The women attended not only to the proper duties of the altar society, but in no small measure to the financial affairs of the church. Fairs and balls were organized and managed by them, the tickets were sold, the collections made and the money put in the bank to the credit of the church.

He might have added that women comprised most of the congregation on a Sunday morning when their men were engaged in that busy day's business. In Protestant churches, they performed the same tasks, as well as serving on church boards, teaching Sunday school, and doing a host of other jobs within the church and the congregation. On occasion, they even did construction or painting.

"Shall We Gather at the River"   103

For all this, women were not allowed in the pulpit. They could sing in the choir and participate in a service on occasion, but they could not be the minister on an everyday basis.

Back east, women would not have been such a dominant force, but in the West they were. They proved they had the ability, intelligence, and stamina to hold these positions and do the work required. They were active beyond the traditional roles of the Victorian wife and mother. Because of their activity in the church and within the community, it is understandable that western states became among the first to give women the right of suffrage. They had proven that they could operate successfully outside the home.

Nor was it unexpected that women, through the church, worked hard for Sunday closing laws and Prohibition. They had become a force to be reckoned with in the mining camps and towns. Granted, some men, and even some women, believed they were going too far and were much too unfeminine in their behavior, but they were fighting a losing battle. There would be no turning back.

In the masculine world of the early mining communities, the church also played the role of social center and outlet—not just for members, but for everybody. Women raised money by preparing dinners that miners, probably tired of their own cooking, thoroughly enjoyed. Ice cream socials were wonderful on a summer evening. One in Ouray, in September 1877, raised $105. Sunday schools also sponsored entertainments featuring local talent and, again, food.

Church social events served other functions as well. Such events were definitely an acceptable place for courting your best girl or beau. One did not need a chaperon at such affairs, as a proper Victorian young couple did when going out together. Also, Sunday school picnics were a great treat for youngsters and oldsters alike.

As might be imagined, the church had to have a proper location. Alex Darley, who was organizing Lake City's church, was offered lots for the church and parsonage. Two of the lots were unsuitable. He wrote, "I had no idea of having our church *killed off by a bad location.*"

Improvisations had to be made in other ways, too. Nothing sounded better than a church bell ringing on a Sunday morning, emphatically stating, as one person thought, "that Jesus Christ had arrived." A Silverton church that had no bell had a member beat "an old circular saw" to call the congregation to services. The Rico paper poked fun at this, saying that the saw made "pure, unadulterated noise" that had no "superior except the calliope."

Even more important for the future of a church in a community was the decision as to whether it would be strictly denominational or a union church. Joseph Pickett, who arrived in Silverton in 1878, preached his first sermon in a schoolhouse to a dozen people who, he wrote proudly, did "excellent" singing. Rather than trying to organize a church, he turned to starting a Sunday school that he hoped would grow into a congregation.

He offered good advice for those considering organizing a church. It was, he wrote, "impractical to sustain several denomination organizations in a small town." He recommended that several groups join together in a church where all Christians could unite on an "evangelical basis." That was probably the right approach, particularly in camps that seemed unlikely to evolve into permanent communities. Denominational loyalties proved strong, however, and jealousies often derailed this practical solution.

Other ministers agreed with Pickett. Fr. John Brinker reached Rico in 1879 and decided it "is too young to sustain churches and ministers." He left, but others would follow, and Rico eventually gained a community church.

The question of whether to build a church always lurked in the background. It was good for community pride but expensive to maintain. As a result, a host of buildings—from schools to homes to court houses to the aforementioned saloon—served as substitutes until the issues of permanency and need were resolved.

By the 1880s, churches had gained a toehold in all major mining towns, and a few struggled in the smaller camps. Ministers often rode a circuit, to preach in schoolhouses, homes, or wherever they could find a room. The congregation might be small, the collection miniscule, but the work was carried on faithfully.

Camps were still being born and dying, so the problems remained the same. Materialism still dominated, although the communities now had more families and women who were pillars of the churches.

Revivals were popular. Prohibition gained momentum, much to the horror of some. Outspoken Gid Propper of the *Telluride News* (February 7, 1885) opined that "Colorado without whisky would be like hell without brimstone." Over in Lake City, meanwhile, the churches and Good Templars were sponsoring a revival that included a performance of that old-time play, *Ten Nights in a Bar Room*. The "Devil is having a hard row to hoe," proclaimed the *Register*. "Down with the devil."

By now, with the churches established, they could even be part of the banter between two rivals. No one was better at this than Ouray's Dave Day and his *Solid Muldoon*. He took on Lake City in early 1883. In his view, "a preacher who thoroughly understands the hidden secrets of draw poker could make a fair living in Lake City. None others need apply."

Money to maintain churches and ministers remained an ever-present problem for the congregations. Telluride's *Republican*, with tongue in

Rico's church today does not indicate the role it once played in the community for the women, families, and reform-minded folks.

cheek, took a poke at this ongoing worry. The collection plate, the article reported, included brass beer checks, brass buttons, and the like. Financial problems troubled all churches. Said one member of his church, "Perhaps a dozen families gave the church loyal and active support," while the rest "were as indifferent as the saloon element."

Not all ministers fit into a mining-town atmosphere. The man had to understand the times, the congregation, and the community situation, but there was nothing unusual about all that. Eben Olcott went to Lake City for church one Sunday in February 1881 and thought the "sermon very was much out of place." He went away "much disappointed" in the minister. The next Sunday, at another church, he heard a "splendid sermon."

Some ministers proved inadequate or, unfortunately, corrupt. Silverton sent one packing after learning that he had amassed bad debts in his previous church and was already running up debts in town.

As the century ended, churches had become traditional in their approach. The Methodist Church in Creede, for example, held preaching services at eleven in the morning and seven thirty in the evening on Sunday, Sunday school at two thirty in the afternoon, and a prayer meeting on Thursday evening at seven thirty. Yet high in the mountains it was much like the 1870s. Harriet Backus lived at the Tomboy Mine from 1906 to 1910. She wrote, "We never saw or heard of a minister going to the mine. Mrs. Driscoll would gather five or six children at the schoolhouse on Sunday and teach them some little Sunday school songs."

Throughout these decades, dedicated Christians worked hard to establish congregations and churches among the materialism and ups and downs of mining. They succeeded in their efforts to the extent of providing an opportunity for people to look beyond their world to the world beyond. The fact that a physical church might not have survived did not mean the congregation failed. For young and old, men and women, the church provided religious, social, educational, and even political outlets. They could ask for no more.

Paraphrasing the popular 1897 hymn "Will There Be Any Stars?", they asked, "Will there be any stars, any stars in my crown when at evening the sun goeth down?" With faith, they believed they found the answer in another song, a 1910 gospel hymn that became famous when the unsinkable *Titanic* went down as the ship's band played "Nearer My God to Thee": "Yet in my dreams I'd be nearer my God to thee." Such faith, solid as a granite-ribbed San Juan mountain, undergirded churchgoers in mining communities.

They were all bound for the promised land sooner or later, and one can hope all San Juaners made it. The words of "In the Sweet By and By" convey their expectations.

> *There's a land that is fairer than day*
> *And by faith we can see it afar,*
> *For the Father waits over the way,*
> *To prepare us a dwelling place there.*
>
> *In the sweet by and by*
> *We shall meet on that beautiful shore.*

Fraternal lodges were found in all San Juan towns and in many smaller camps. They provided the newcomer a door into the community and a gathering place for old-timers.

CHAPTER NINE

Age of Joiners

*The first question for the debate before the
Silverton Literary Society was resolved that a burro
has no rights which a man is bound to respect.*
—Lake City Silver World, October 16, 1875

*The Devil is having a hard row to hoe in Lake City
just now between the churches and Good Templars.
Down with the Devil.*
—Lake City Mining Register, December 7, 1883

*The Chippewa Tribe of the Red Men entertained a large
number of pale faces and ladies of Pocahontas at their
wigwam in the Knights of Pythias Hall.*
—Telluride San Miguel Examiner, April 27, 1907

Nineteenth-century Americans, whether San Juaners or anyone else, were joiners. There were innumerable fraternal lodges, clubs, and other organizations San Juaners could join, and they did so with an enthusiasm their children and grandchildren seldom matched.

Telluride, for instance, according to the *Journal* (December 30, 1899), had the following fraternal orders:

Woodmen of the World, with 112 members
Ancient Order of United Workmen, 70 members
Order of Improved Red Men, 88 members
Daughters of Pocahontas—auxiliary order of the Order of
    Improved Red Men, 53 members

> Forresters of America, no member total listed
> Knights of Pythias, 106 members
> Bethany Temple 25 (Knights of Pythias), 50 members (men and women)
> Telluride Lodge No. 56 A.F. & A.M., 114 members
> Telluride Royal Arch Masons, no member total listed
> Knights Templar, 30 members
> Order of Eastern Star, no member total listed
> Independent Order of Odd Fellows, 117 members
> Sisters of Rebekah, 88 members

The paper added that there was "not probably any city in Colorado outside of Denver" that had such "luxuriously and commodiously housed" Masonic lodges.

Declining Rico, by contrast, offered the Masons, Odd Fellows, Woodmen of the World, and the Grand Army of the Republic. Apparently, the San Juans never had chapters of such exotic organizations as the Hoo-Hoos, White Rats, and Gu Gu, for none were reported.

Even considering that joiners, both men and women, were often members of more than one lodge, these figures still represented a healthy percentage of Telluride residents. The 1900 census takers counted 2,446 residents of the town. Even including the children, these lodge members represented 34 percent of the population.

Add to this the members of the Grand Army of the Republic, social clubs, and such groups as the Western Federation of Miners, and Telluride certainly seems to have been well organized. This was not unusual; even the smallest camps generally had one of the four most popular lodges—Masons, Odd Fellows, Ancient Order of United Workmen, and Knights of Pythias. The larger the community, the more lodges it had, as occurred in Telluride.

The number of lodges people could join was truly amazing, and join they did. Interestingly, the 1880s through the turn of the century represented the absolute peak in the number of new groups being formed in the United States and in the total number of organizations, which extended well into the hundreds.

The reasons for organizing these groups and the functions they served varied considerably. They certainly provided continuity and a sense of belonging for their members as they drifted throughout the San Juans from community to community over the years. Each lodge provided social activities for its members, such as meeting places, dances, parties, and dinners. Members might make new friends and business contacts, learn about the community, or gain insights into current local events. The groups offered

How important were clubs and lodges? Silverton's San Juan Democrat *counted more than half a dozen in 1888 that held regular meetings; in 1899, Telluride had seventeen.*

association with men (most of the lodges were limited to male members) with similar cultural backgrounds and perhaps political affiliations. All these offered keys to a new home.

The fraternal lodges were benevolent and even beneficiary organizations, aiding less fortunate members and their families. Lodge cemeteries were organized and maintained for the brotherhood to find a final resting place. Living members, it was hoped, would support the grieving family.

Special life insurance policies for members provided another inducement to join. In the Ancient Order of United Workmen, for example, each man upon joining received a beneficiary policy of two thousand dollars, which cost an average of sixteen dollars per year. The dangerous work a large proportion of the population engaged in made it almost mandatory for a family man to provide some assurance of aid for his wife and children in case an accident befell him. Many insurance companies would not insure miners or others in risky occupations.

Decoration Day was a big event for the members of the Grand Army of the Republic. By the 1890s, "the sympathies of the soldiers are the same whether north or south."

In that time of fly-by-night insurance companies and the high rates respectable firms assessed, the fraternal societies provided a safe and perhaps the only economical substitute. One could count on his brothers, not some unknown stranger, to fulfill the obligations of the association. Members turned out for a deceased member's funeral and gave financial support and comfort to their families, as well as a place to turn if a need arose in the future. Where else could miners, merchants, and workers secure social and benevolent benefits for so little money?

Besides the lodge meetings, which gave members one or two nights out each month, social events the lodges sponsored provided another get-together outlet for husbands and wives and often their children. For years, the Chippewa Tribe of the Improved Order of Red Men hosted a Thanksgiving evening oyster supper and masquerade ball. Telluride's *San Miguel Examiner* hailed the November 1906 event as "one of the biggest dances of the season" adding that the supper offered oysters "served any style." The same "tribe" the next April held a razzle-dazzle evening featuring card playing, a "splendid lunch," and dancing, although the paper added that only the young people lingered long at the dance.

The Knights of Pythias sponsored an annual masked ball in Ouray in the 1880s and 1890s. Even the installation of officers might be a public affair, as it was in Creede. The local banjo and guitar club, a piano solo, and an "able address on the subject of Phyhianism history" followed by other music and several more speeches concluded the January 1893 evening. The audience, the reporter remarked, "was well pleased and passed the evening enjoyably." The Masons usually held a grand ball on Washington's Birthday, while others might select Valentine's Day, and the Christmas season provided the occasion for various lodge parties. The week between Christmas and New Year's Day usually found more miners in town than any other time, except the Fourth of July, for those were the times the mines generally closed down.

The Grand Army of the Republic (GAR) emerged as probably the most universal organization in the San Juans, at least until the number of veterans decreased. With all Union veterans eligible to join, each of the large towns had camps, which members from smaller communities might also join. The GAR represented more than just a social group; it politically supported the Republican Party and lobbied continually for pensions for members. Old soldiers also reminded younger Americans about those traumatic war years from 1861 to 1865 and the youth they and their comrades had given up to save the Union.

They also could be counted for Decoration Day, as Memorial Day was called then. A ceremony for departed comrades at the cemetery, along with a patriotic oration and perhaps a short parade, reminded old and young alike of the sacrifices the boys in blue had made during the War between the States. By the 1890s war ardor had cooled, and former Confederates joined the parade in Creede and elsewhere, along with fraternal organizations, schoolchildren, and citizens in carriages. Passions had indeed abated, if one reporter's remarks represented his readers' feelings. "The sympathies of the old soldiers are the same, whether from the north or from the south. Their sufferings and dangers were the same. It was a war in America by Americans, a struggle over principles in which the people differed."

When a community held a parade, the lodges and the GAR usually turned out in full uniform, forming a significant portion of the marching line. Eventually, though, when the aging veterans found marching too difficult, they rode in carriages, continuing to remind watchers of the days of '61.

Other short-lived associations also flourished, often for a single purpose. Lake City in 1878, for instance, had a Miners' Library Association that helped start the town's library—in this case in rooms, not a library building—and purchased books. Occasionally an association sponsored a lecture or a lecture series. One speaker grabbed the irascible Dave Day's attention in Ouray, and she caught the full blast of his editorial pen. "When Miss Kate Field, or any other lecturer, strikes Ouray for the purpose of unloading a lot of sentimental rot regarding Dickens or others she, her, him, or they must prepare for honest, frank and just criticism."

During the slow winter months, various groups appeared on the scene, some serious, some not. As early as December 1875, Silverton's Literary Society debated the merits of one earthshaking question: "Resolved that a burro has no rights which a man is bound to respect." Ouray, however, had a serious literary society, according to the Reverend James Gibbons, who helped organize it. While it only had a few members, the aim was "seriously given to self-improvement" by discussing moral and social questions. That Gibbons remembered the issues occasionally caused "lively debates." Rico's literary society not only discussed issues but occasionally held a musical evening.

Lake City had a Debating Society in the winter of 1883. They discussed such stirring issues as "whether or not the federal government should own and control the telegraphic system of the country." Despite the subject, the December 7, 1883, newspaper report claimed that "the debaters straddled

"The Firemen's Masquerade Ball was the event of the winter season," thought a Lake City resident. The fire laddies wore their badges with pride.

each other" and the referee "found it almost impossible to keep them within bounds." It turned out to be "quite amusing" entertainment.

With the evenings long and cold at Animas Forks in 1882, locals organized a social club with dues and officers. Among the events they sponsored was a September ball with a "delicious" supper and "pretty good music." Couples came from as far away as Rose's Cabin, Capitol City, Burrows Park, and Silverton. A year later, with the social club apparently having gone south, a Bachelor's Club was created to get the men through winter's "terribly stiff weather."

Apparently, some of the gay young blades, and older ones as well, faced challenges on the dance floor, finding it hard to trip the light fantastic. Thus, it was not uncommon to have dancing schools, which some "professor" taught, to help pass the long winter evening.

> **OURAY LODGE**
> **Societa di Giorge Washington**
> OURAY, COLORADO
>
> ther Louie Fedel under the obligation of the order oses for membership in this Society, Mr. Dominick Mattivi
> Signature of proposer X *Louis Fedel*
> tate your name Dominic Mattivi   Residence Ouray Colo.
> ccupation Laborer   M. S. W.   Single
> ace and date of birth Regnan Italy
> foreign born, when and where were final naturalization papers issued Came in this y 1930 and his father was a Naturalized Citizen
> you know of any physical ailment that might cause you to become a burden upon this or No
> ames of three members of this order as references: Eggio Delpaz, William Groff, & John Casnagranda.
> QUESTIONS MUST BE ANSWERED BY THE APPLICANT

Lodges served a variety of purposes, including remembering the old country. They also gave pride to the community; Telluride claimed its lodge halls were "luxurious."

In addition to the lodges, associations, and clubs, a variety of other sociofraternal clubs appeared on the scene. The Good Templars, humorously referred to as "water tanks," served the cause of temperance. Both men and women of good standing could join. For women more serious about the liquor problems, the Women's Christian Temperance Union and other prohibition groups attracted their attention. Where there were enough foreigners, a club such as the German Turnverein Society held occasional meetings.

Churches also sponsored groups, particularly the Good Templars, children's groups, women's societies, and "antigroups"—antigambling, antiliquor, and antiprostitution—as well as serving as the meeting place for lectures. They were also patrons of social affairs, but the church's negative attitude toward dancing often limited attendance. The fire laddies also sponsored dances, fund-raisers, and other social events, including the ever-popular fire company races and games. Town baseball teams were only a step behind them in offering events.

The controversial Western Federation of Miners (WFM) became popular in the San Juans as management's grip on the miners tightened in the 1890s. Besides representing labor, the group served a social function, hosting dinners, picnics, and dances. It also underwrote the cost of a hospital in Telluride. With the strikes of 1903–4, however, the WFM disappeared from the local scene.

Each of these organizations had a stake in the mining camps, and each endeavored to improve the community. By providing bonds of fellowship in a transitory life, they not only served as havens for members, but could also coalesce on behalf of a variety of issues, including law and order, civil improvement, and creating a lasting community.

These diverse groups provided a vital element in the mining community for the reasons discussed. For many members, they also bestowed a sense of importance as "brothers" moved up the officer ranks to become a chief Hoo-Hoo or whatever or gained another significant role. In the often-impersonal life of these towns and camps, this could be a badge of acceptance for an otherwise ordinary individual.

Without these groups, life in the San Juan mining communities would have been much duller, less sociable, less caring, and less like the homes these folks had left behind. They also provided a safety net in a world not known for offering one. For all these reasons, each and every organization made its own contribution.

What was a saloon without a nude painting gracing a wall? The red-light district included saloons, low-class variety theaters, gambling "hells," and prostitutes.

CHAPTER TEN

Sin, Sex, and Leisure-Time Pleasures

This part of the camp we well named "Hell's Acre," for the first part of the name was about all that was ever raised in that acre. There was always a sad thought in my mind connected with this portion of our camp; i.e., that so many young men who had been well trained in eastern homes would visit the dance-hall to see something of "wild life" in a frontier mining-camp during its palmy days.
—George Darley, *Pioneering in the San Juan*

Another gone wrong in life, over whose acts we draw the mantle of charity, hoping before Him who rules she may find the forgiveness for her fall on earth for which she prayed with her dying breath.
—Morning Durango Herald, March 29, 1887

Dance halls are the product of new mining camps in the mountains or boom towns on the border of civilization, and Ouray has long since outlived the excuse for such institutions, if indeed, there ever was any.
—Ouray Herald, May 9, 1902

Such sentiments were often expressed about the red-light district and its denizens during the heyday of San Juan mining towns and camps. The district, the gambling, the life-style, and the gamblers and prostitutes repulsed and fascinated both their neighbors and visitors.

Damned and praised, tolerated and hated, sinful and profitable, segregated and wide-open—but never ignored—mining communities' red-light districts were lucrative ventures. In the male-dominated world of mining, entertainment focused on the district, and that brought money into the community.

Legendary, thanks to fiction and Hollywood, the red-light districts supposedly featured cribs, parlor houses (not to mention pretty prostitutes and madams), and saloons with stage shows, beautiful bars, and pretty waiter girls far fancier than most men's home-away-from-home establishments. In reality what was in the districts included, at their best and their worst, high- and low-class saloons, gambling "hells," variety theaters, parlor houses, and cribs.

For the growing Victorian middle class, all this represented sin in capital letters. At best, the district had to be tolerated in its own section of town and as far out of sight as possible. At worst, it threatened the community, tempted youngsters, and debased the men and women who indulged. Appalled reformers, shocked that city fathers allowed such degradation, lewdness, and sinfulness to exist and flourish, sallied forth to bring repentance.

Some accused the districts of being centers of drunkenness, fighting, debauchery, crime, and drugs, which in a sense they were, unless heavily policed. Yet they were found everywhere, and a community without one seemed poor in the viewpoint of many locals. It became almost a badge of success and prosperity to have a more flourishing red-light district than one's neighbor, at least, that is, until the reformers finally gained the upper hand.

No respectable Victorian women would be caught dead in the district; they would make a point to walk on the other side of the street to avoid such "contamination." They might storm into a saloon to haul hubby out, or a very few might walk discreetly down a back alley to enter a sampling room at the saloon's rear and have a drink, but otherwise they ignored such shameful carryings-on.

For the business community particularly, but others as well, the red-light district represented an attraction that brought customers into the camp or town. Without it, they might go elsewhere with their business; thus, the red-light district became, and remained, a vital part of the community.

Numerous tourists, however, came to the mining West to see and even sample such sin and goings-on. Upright folks who back home would not be caught dead in a saloon or wandering about the red-light district (if there was one) where their shocked neighbors might see them sampling some of the pleasures socially frowned upon in their Victorian society. Mining-camp editors chuckled at the hypocrisy.

Eventually, though, as the mines and communities declined and the easy money faded away, the drifting crowds wandered off to more promising locations. By then, too, respectability had gained the upper hand and the era ended, although saloons hung on until prohibition won the day in Colorado in 1916.

"Three female divines—Long Annie, Mollie Foley, and Lizzie Gaylor—were among the distinguished arrivals this morning." Three of Telluride's cribs grace Pacific Avenue.

For now, however, miners had money to spend, or so it was thought, and some folks soon arrived who wanted to make sure that money did not languish in their pockets for too long. These opportunists came on the heels of the first prospectors and urban development in the 1870s. Before they left, the entire mining era had passed into history.

Lake City, one of the earliest San Juan mining towns, illustrated how quickly all this transpired. If offered a variety of temptations beckoning footloose young men in and around the town and coming in from the nearby mines to let off some steam, as they expressed it. Its pioneer newspaper, the *Silver World*, reported (October 23, 1875) that "impromptu entertainment gotten up in the City Saloon last Monday developed some very fine minstrel talent."

Sin, Sex, and Leisure-Time Pleasures

The next year, 1876, the editor noted that three-card and Spanish monte were being played in town, warning readers that such games would "attract the unwary and fleece them of their shekels." By 1878 a "beer garden" was flourishing with a "large platform built for dancing and music furnished by the management."

That same year, the American House Saloon advertised that every Saturday night, from eight to midnight, a "free lunch" would be served and "all were invited." Tired of their own cooking, few bachelors, or perhaps a stray married man, could resist a free lunch, which featured salty items that would help enliven the thirst for beer and whisky.

The fallen women, fair but frail, erring sisters, brides of the multitude, female divines, and ladies of the evening, euphemisms used to avoid those shocking terms prostitute or whore, fascinated people either in a moralistic Victorian way or as a lustier attraction. While most people looked upon the women as degraded and of little good, others did not.

One of the latter was writer Frank Fossett who described prostitutes in his 1878 book, *Colorado: Its Gold and Silver Mines*. His approach, despite being staunchly Victorian, raised some interesting points:

> To the abandoned and fallen women, who followed the army of prospectors like a swarm of locusts, some credit is due. Wretched and degraded, ignored by the Christian, and spat upon by the so-called moralist, victims of a jeering world, they forgot "Man's inhumanity to man" and were the direct means of saving many valuable lives from that dreaded disease, mountain fever.

Thus appears in one of the earliest Colorado and San Juan manifestations the prostitute with a legendary heart of gold.

Occasionally a San Juan story appeared that did lend some credence to that image. Dave Day published one in his *Solid Muldoon* (September 20, 1889) under the headline, "Deserted and Friendless." A sick, pregnant woman, whose husband deserted her along with their children, was nursed and helped by a sporting-house landlady.

The prostitutes, like many others, drifted in and out of the San Juans. For them, the best season in the high mountains and smaller camps came during the busy summer months. One of these women was that "notorious prostitute and thief Bronco Lou." She, like many of her contemporaries, arrived in Silverton in the summer and left as the snow started to fall. The *La Plata Miner* breathed a sigh of relief when she was lynched in New Mexico in late summer of 1881.

The *Telluride Republican* (July 3, 1886), for one, did not like people taking advantage of the women. When a miner engaged a room at the Watson House and "surreptitiously occupied it overnight with a frail sister from one of the sporting houses," the editor reported, "he had the gall to walk out without paying the bill." The woman or the house or both came up short initially, but the dastardly villain was caught eventually and fined. The paper remained silent on whether it was more concerned about the loss of a night's room rent or the girl's lost pay or the sin.

By city ordinance or public pressure, the women were forced to live in certain areas of town. If they left that district, the public generally took up arms. Silverton had such a scandal in late 1884 when a "disreputable woman" rented a house and moved into a "respectable" neighborhood. Her neighbors "were greatly scandalized by the behavior of this disreputable woman." It took a while, but this "disgrace to the fair name of Silverton" was finally removed.

Still, having a number of soiled doves in a red-light district was looked upon by many as a positive development. When a "herd" arrived in Ophir, the nearby Telluride newspaper observed that their arrival "is the best sign of money being made there." In Ouray's Dave Day's description, when a group arrived there, it "filled the bill of the village."

The prostitute's life was not to be glamorous, despite some pretty trappings and the sentimentality of later writers. Suicide was a continuing problem, particularly around Christmastime, when thoughts of home and earlier holidays came to the fore in the women's minds. Dave Day reported on one such case in January 15, 1886. "One of the despondent daughters of prosperity took a fill of blues last night and with it a dose of cold poison in the shape of morphine." After "heroic doses" of mustard were administered, the attending physician brought her "back to this practical world. At last reports she had braced up and was willing to remain with us a while longer."

For both the women and the customers, a risk existed despite taking precautions. There was no known cure for syphilis, gonorrhea, or other sexually transmitted diseases. Patent medicines cured the symptoms but not the disease.

Who were these women who worked on the line? Ida, Lou, Edna, Belle, Mattie, Mollie, Hattie, and the others are best identified in the census returns. The 1880 federal census taker found eight in Lake City. They were all native-born, in their twenties, and single, except one who listed herself as widowed. Four came from the Midwest, two from the East, and one each from the South and West.

Mark Twain observed that the "chief gambler, and the saloon-keeper, occupied the same level in society, and it was the highest." This Dewey Musical Cabinet was a nickel slot machine.

Nearly the same pattern appeared in the 1885 state census for Ouray. Four of the thirteen women were nineteen, the rest in their twenties. Two were widowed, two divorced, and the remaining nine single. Two came from England, and the others were native-born Americans. These sporting women lived in a definite section of town in two groups, but there was no indication whether one might have been high class and the other lower class. That depended on a woman's age and attractiveness. Sometimes class was determined by fads; red-haired Jewish women, for example, were thought to be sexier.

It is difficult from the raw census returns to find out much about these women, their education, skills, background, or anything else that might have pointed them toward prostitution. In the Victorian scheme of things, a divorced woman, for instance, was looked upon as having failed as a wife. A double standard existed: men could sow their wild oats, but women were put on a pedestal from which they might fall, to the horror of their respectable counterparts.

It was hard for them to get off the line; the stigma proved too great. A few married. According to one old-timer, some miners "married those women—they made wonderful wives." Mining engineer Robert Livermore remembered one prostitute at Telluride. He and his wife were looking for a "maid of all work" and found one. "We settled on a lady with a past, a graduate of the 'row' who must have had attraction, as she subsequently married twice. She was with us for much of our stay in Telluride, an excellent cook and an exemplary character."

Dance halls were sometimes places to meet ladies of the night. The description of one in Telluride in the early 1900s presents a fairly typical example. There was a long bar on one side of the room, with a "fence separating the bar from the dance floor." The band sat in a corner on "an elevated platform." It consisted of three or four pieces, "always a piano" and perhaps a bass drum, cornet, and fiddle. The girls sat along the dance floor on a bench.

According to one man's memory, they wore "knee-length skirts with low-necked dresses, but not so very low—just enough exposed to display a string of beads."

A dance lasted two to three minutes, then the man "was expected to lead [his] girl to the bar and slap down 50 cents. You get a drink and the girl a brass check."

Sin, Sex, and Leisure-Time Pleasures

"Belly up to the bar boys!" Many a congenial evening was spent drinking and talking about the "man's home away from home."

Silverton's Ernie Hoffman remembered the dance halls. "Each one had at least a dozen girls and a big bar out front." When the girls saw you coming they would "grab you." He felt the "dance didn't amount to a damn. It wasn't very long when the piano player quit playing and you had to go to the bar to buy a drink ($1) and a dollar for the gal, she drank tea, you drank booze."

Ouray, in May 1902, had finally had enough of the dance halls. In a wave of reform, the city fathers closed those halls "of music where is liquor sold." Saloons and sporting houses promptly stopped their musical entertainment. This sudden emergence of civic virtue upset Telluride. The *Journal* said of the closing of the Ouray dance halls that "it dumped a lot of superannuated fairies and nondescript musicians into Telluride." In contrast the *Ouray Herald*, (May 9, 1902), praised the reform, which "marks the beginning of a new and more progressive era in the social conditions of Ouray."

The editor addressed the changing times, as quoted at the start of this chapter. He did not like the fact that the tenderloin was located near the railroad depot and "all passengers in and out of Ouray are compelled to pass near the vicinity." This gave Ouray a black eye because the girls lined up "along the sidewalk in front of the halls when the trains arrive," giving strangers a "bad impression of the condition of society" in the city. Visitors and tourism were emerging as important economic pillars, and a negative initial image did the town no good.

Low-class variety theaters served as a combination saloon, gambling hall, and dance hall, along with offering some type of variety show. Hollywood has overplayed this part of the western story; the theaters were not overly common in the San Juans. A few could be found in the larger towns within the red-light district.

The shows were presented only to advertise the girls or lure men in to take part in the more profitable gambling and drinking attractions. They had been more common in earlier mining rushes, particularly in California. The only difference in the physical appearance of the theater from the dance halls described earlier was a stage and perhaps boxes where the pretty waiter girls could entertain customers. The Gold Belt in Ouray offered these attractions until the infamous city-council decision curbed its attractions.

Gambling was pervasive in the mining towns and camps. The miners put themselves in danger every time they went underground, so taking risks at the card table or roulette wheel was consonant with their lives. Again, to turn to Robert Livermore:

> The town [Telluride] was wide open. I believe there were thirty-odd saloons, in most of which a roulette wheel and faro game were established. The feeling among the mine operators was that the sooner the miners got broke, the sooner they went back to work.

Ernie Hoffman, who grew up in Silverton, discussed gambling with the author. Most of the saloons had a gambling license for, "damn near anything, any kind of a damned gambling game you wanted."

Drinking might lead to alcoholism, gambling might develop into an addiction, and prostitution could cause an incurable disease. Some men fell victim to one or another or even all of them, creating individual, family, and community problems.

Drugs also became a problem. Opium, belladonna, cocaine, and other drugs could be purchased over the counter at drugstores. The problem of addiction was a new one that finally came to be realized in the 1890s, but no solution appeared except to advocate either going cold turkey, drinking a host of patent medicines, or committing yourself to a Keeley Institute.

A madam controlled a parlor house, taking a percentage of each trick and providing protection, among other things, for her girls. Telluride's has seen better days.

The latter promised to cure the liquor, opium, morphine, and tobacco "diseases." Durango had such an institute, which San Juaners who fell victim to addiction entered. Unfortunately, Keeley's long-term success failed to equal its promises.

Too often popular patent medicines, which promised to cure just about every aliment, tried to cure opium addicts by substituting some other addictive drug. Until the passage of the Pure Food and Drug Act of 1906, what might be in a bottle of this or any cure-all was anyone's guess.

Another problem was what Victorians beguilingly called social diseases. As mentioned, no known cure existed for venereal disease, just medicine that treated the symptoms. Still, as long as prostitution was allowed, the problem would exist. Silverton's city government and those elsewhere tried to resolve the issue by having the women on the line undergo a medical examination when they came in to pay their monthly fines. That might have eased customers' minds, but, as a doctor pointed out, the certificate on the wall was only good for the first customer even if precautions were taken.

City governments were in a quandary in trying to decide what to do about red-light districts. They needed them to attract men to town and

Ouray's Gold Belt dance hall had also seen better days. In its heyday, however, it was flourishing part of the red-light district.

stimulate business, but at the same time, they could become social and perchance criminal problems. The red-light district also provided a source of government revenue as the saloons, games, and girls had to get licenses or pay a monthly fine to continue to operate. Whether that balanced the cost to city government of policing the district raises an interesting but unanswerable question.

Saloon keepers in Lake City, for instance, had to pay $125 for a three-month license or $500 per year, while a livery stable paid $20. Creede's license was $600, and it had a fine of $10 for keeping a "common, ill-governed or disorderly house." Telluride placed the fine at between $10 and $25. Monthly fines followed as the city fathers winked at such law breaking.

City councils or boards were, in fact, rather busy passing ordinances referring to the districts. "Females were not to loiter" about saloons, grocery stores, or any "public place where liquor is sold or exchanged or given away" in Telluride. Rico passed an ordinance proclaiming that "keepers of all bawdy houses shall be held responsible for the disorderly conduct and acts of public indecency of the inmates." Ouray defined "decoying females" and pimps as vagrants, and they were treated accordingly. Drunks in Silverton could be arrested with or without a warrant; "lascivious living" was banned everywhere and usually involved a man living with a "prostitute, courtesan, or lewd woman."

Obviously a double standard existed. Men involved with the fair but frail were not fined or imprisoned. Councils rarely targeted them unless they sold liquor to a minor or appeared in a state of drunkenness.

Victorians, like parents in other generations, worried about the influence the district might have on children. As a result, city governments voted for a variety of ordinances. Rico passed one prohibiting minors from "frequenting saloons," with a license revocation if violated. The city also ordered placards to be placed around town to notify one and all about the ordinance. Anyone younger than twelve in Lake City could not shoot pool or frequent a poolroom. Telluride made the age fifteen, and Ouray and Silverton set it at eighteen.

An ordinance establishing a curfew for children appeared to be one of the most popular means of removing them from alluring temptations. Creede established a winter curfew of nine at night and a nine-thirty summer one for those under sixteen. It was nine every night in Ouray year-round. Selling liquor to a minor was an offense in all communities, with fines ranging up to twenty-five dollars.

As the years slipped by, the red-light districts declined as communities settled down and mining waned. Reform was in the air by the time World War I broke out in Europe in 1914. Also, by then the remaining towns were enforcing ordinances that may have been simply on the books before, but that were honored more by the lack of enforcement. Prohibition in Colorado in 1916 eliminated the saloons, and the decline of mining took away many of the customers of parlor houses and cribs. Dance halls had long since disappeared, and gambling was looked upon by many as a social problem and a community disgrace. Along with drinking, it was driven off Main Street.

The old Nevada silver miner, Mark Twain, had fortunately passed from the scene by that time. He, like a lot of miners, enjoyed his drink. Thinking about whiskey as he reminisced about his steamboating days in his book *Life on the Mississippi*, he described the early San Juans. "How solemn and beautiful is the thought, that the earliest pioneer of civilization, the van-leader of civilization, is never the steamboat, never the railroad, never the newspaper, never the Sabbath-school, never the missionary—but always whiskey! . . . Westward the Jug of Empire takes its way."

A town band was almost a necessity for parades, summer-evening concerts, dances, and an occasional chivaree (joined by friends with noisemakers) to serenade a newly married couple.

CHAPTER ELEVEN

Culture Arrives in the San Juans

*To me an opera is the very climax & cap-stone of the absurd,
the fantastic the unjustifiable. I hate the very name of opera—
partly because of the nights of suffering I have endured in its presence.*
—Mark Twain, *Mark Twain:
His Words, Wit and Wisdom*

*Nothing can make a Wagner opera absolutely perfect and
satisfactory to the untutored but to leave out the vocal parts.
I wish I could see a Wagner opera done in pantomime once.*
—Mark Twain, *Mark Twain:
His Words, Wit and Wisdom*

*It's day all day, in the daytime,
And there is no night in Creede.*
—Cy Warman, "Creede"

San Juaners might have agreed with their more famous mining contemporary, but that did not limit their desire to have in their communities the capstone of nineteenth-century culture—an opera house. Opera was likely never performed there, but just having the building meant a town had come of age culturally. An opera house, however, proved too much for camps to attain.

If a mining community and its citizens ever hoped to emulate their eastern colleagues and society, an opera house became a must. So were theatrical performances, Chautauqua speakers, literary societies, and a host of other events that defined Victorian culture beyond the home.

Creede's city government charged "traveling theater and troops and exhibitions" ten dollars per night for a license. Circuses paid fifty dollars per day. Telluride gave locals a break at five dollars.

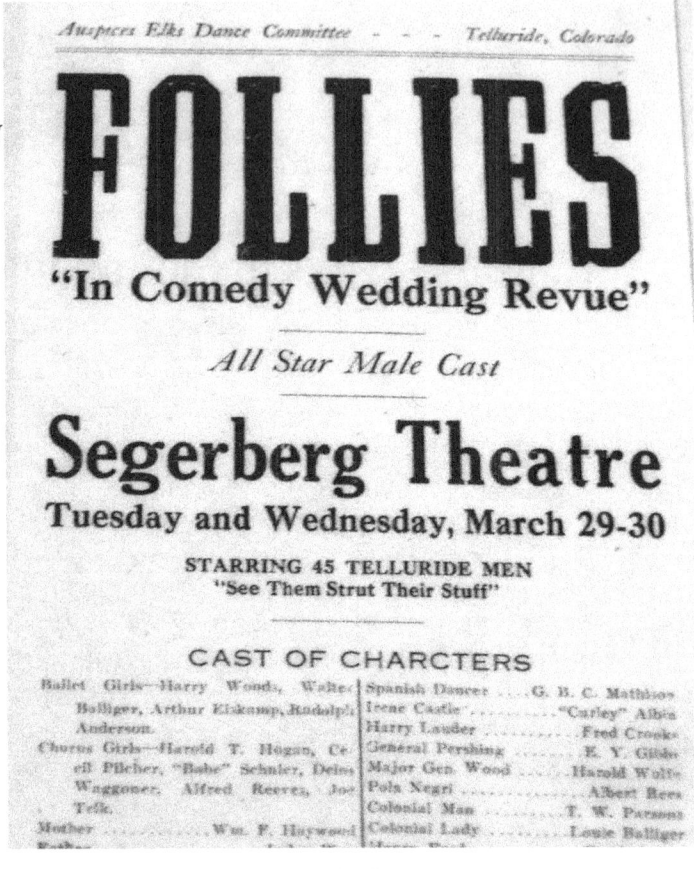

Ouray, Creede, and Telluride did build opera houses; Durango, Silverton, and the rest of the towns in the San Juans did not. Although some claimed other buildings to be an opera house, no San Juan structure had the architectural grandeur inside or out of the two silver-millionaire Horace Tabor built for Leadville and Denver. The lack of such a majestic edifice hurt locals' self-image, but they carried on, for no Tabor came forth to bestow this cultural gem on their cities.

Durangoans pleaded for an "enterprising capitalist" to build an opera house. What the *Herald* envisioned in July 1894 was a "first class opera house" with a seating capacity of one thousand. Although not of that magnitude, the town had a theater locals proudly called the Durango Grand Opera House.

Often, the first appearance of culture was a group organizing for some uplifting purpose. Lake City, for instance, had a library association as early as 1878, and Ouray and Rico acquired literary societies soon thereafter. While these groups were more active in the winter months, they provided a year-round, sounding-board involvement in cultural activities. The public

Every mining community desired having an opera house, even if an opera never graced its stage. The appearance of culture was as important as its substance.

schools helped as well by acquainting young people with culture, plays, and concerts before proud parents and other locals. Women stood in the forefront in attaining many of these developments.

Libraries could also serve as gathering places for those seeking something beyond the average mining-camp or town life-style. In its July 10, 1886, issue, the *Animas Forks Pioneer* praised Silverton's public library and reading-room association. "Too much credit can not be given to the ladies who successfully started and carried out the movement for a free reading room." The editor especially praised the books in this "well selected library."

Not everything was peace and quiet in the library, however. Silverton's *San Juan* complained in April 1887 that the "library is no place for small boys to loaf and play." The library was no "play house," and the editor recommended the "the marshal would do well to keep an eye on small boys in the habit of frequenting the place." Where was the librarian during all this?

Library, yes, theatrical performances, definitely. In 1909, thanks to the help "of a number of our best people" who "guaranteed money to pay for the attractions," Telluride secured a series of theatrical performances. The Congregational Church had sponsored a "course some years ago," but it made "little or no money." Times had changed, the *Examiner* believed. Under the new opera house management, it now promised wide success.

The shows were promised to be the best money could buy, and people had to purchase tickets because the performers would not play "unless paid before they went on stage," the *Examiner* warned. The coming attractions featured a wide variety, something certain to appeal to almost anyone. They included lectures, a magician, college-age singing girls, a "select musical

For those who enjoyed a concert in the park, Silverton provided a bandstand. The town band, along with the baseball team, were symbols of local pride.

organization," and a New Zealand native family. This last show promised to have "moving pictures and curios" as part of the performance and exhibit.

Apparently the shows did not always draw a crowd. In January 1907 the Telluride opera house announced that "prices have been reduced" for a week-long series of plays by the "Glenn Company." The opera house needed "public patronage" sufficient to justify bringing this outstanding company to town.

San Juan reviewers could be critical of the performances as their eastern counterparts. For example, the *Silverton Standard* (January 23, 1903) opened a review with the comment that "'The Drummer Boy of the Rappahannock' has come and gone, but leaves no aching void." The reviewer added that the company "did not meet expectations or promises and the work was a decided disappointment."

A few years later, Telluride's *Examiner* (April 20, 1907) reported that "The Two Orphans" was "very well rendered." Unfortunately, *The Scout's Revenge* proved an ordinary border drama. Nor did the appearance of Miss Howard, a spiritualist, appeal to Durangoans. "The affair was a fake," complained the *Herald*. The "very ordinary" performance did not sit well with Durangoans, and "most of those who attended were kicking themselves yesterday."

For miners and others living at the mines high above these San Juan communities, a trip into town was not an everyday event and probably rarely, if ever, one made for cultural reasons. Harriet Backus, who lived and wrote about her life at the Tomboy Mine, remembered going to Telluride three times, none of which was "to attend a play."

Something new appeared onstage after the century's turn—vaudeville. Creede proclaimed it "the favorite entertainment theatrically nowadays," and Collins Opera House had "eight big acts." Telluride concurred, noting that "vaudeville now so popular all over the country" can be seen at the opera house "for only three more nights." In these variety shows, some act or individual was bound to please each member of the audience.

The new wonder of the age, the silent movie featuring a piano player to set the mood, soon appeared. Despite a "heavy rain that kept the crowd down," those Silvertonians who attended *When the Earth Trembled* in August 1914 found it "pathetic and thrilling." Indeed, the earth seemed to tremble that month as war broke out in Europe.

The year before Creede's "motion picture theatre" had promised movies three times a week with "three reels at every show." All this for only ten cents! Durango had three motion-picture theaters by 1910 to lead the

San Juans. It also had, just before war broke out, its own Durango Film Producing Company.

Few towns did not have a town band, perhaps the most famous of which was Silverton's Cowboy Band. The *Standard* (May 3, 1890) proclaimed it "the coming band of the West," something the editor thought would "be great advertisement for Silverton and the San Juans." The band had sent to Texas for "two pairs of the largest cow horns to be had" and was "engaged for a trip to Denver."

The bands played for holidays, concerts, serenades, parades, and other musical events. Much like a baseball team, they represented the community and, in their case, its cultural attainments. They might occasionally perform at a dance, although dance bands usually had fewer members.

Creede had both a band and an orchestra, something of which the town "may be well proud." The two probably had relatively few members, and were likely made up of many of the same people, but in the local estimation, they were "superior to many like musical organizations of larger size." As the proud *Candle* (February 1906) implored its readers, "We sincerely hope citizens give the boys every encouragement to keep the organization going."

Singing groups also appeared, in addition to church choirs. The Revenue Mine sponsored a minstrel company that "entertained during some long winter evenings" in 1898. The *Ouray Herald* praised the group for having "far above average talent." Lake City singers organized a glee club to provide entertainment and to "be an object of pleasure" for the participants during the cold winter months of 1896. Alex Botkin, who worked at the Tomboy Mine after the turn of the century, remembered the mine's Cornish choir. "They had a very good choir and took great pride in it."

By 1900 San Juaners were gaining enough collective memory of things past to start reminiscing, remembering, and writing. Also, some had enough leisure time and desire to become more involved in other cultural activities, such as painting and writing poetry. It turned out to be a very small cultural blossoming for this mining district and its communities.

Throughout the period under discussion, a few San Juaners dabbled in painting, but few equaled the outburst of activity at Ophir in 1909. The women of the community became involved in a variety of artistic endeavors. One was working on hand-painted plates and "meeting with unlimited success." A reporter enthusiastically proclaimed that "her columbines are as true as the flower itself." Another woman conducted a class in painting china and "the five ladies [who took the class] have beautiful china to show for their summer's work." Finally, "Miss Ruth Long, one of the most talented young ladies of Ophir, finished some very beautiful water colors."

This newfangled machine actually played music. So popular was it that concerts were played over telephone lines for all to hear.

As the new century dawned, a poet and a novelist began to chronicle the saga of a mining region and an era. Both had been actively involved in mining in the San Juans. They were not alone in using a mining background for literary efforts. They walked in the footsteps of Mark Twain, Mary Hallock Foote, Bret Harte, and Robert Service.

Mining engineer Frank Nason wrote two novels about the San Juans. During the years 1898–1900, he served two years as geologist and mining engineer for the Mount Wilson Gold and Silver Mining Company, operating the Silver Pick Mine. Located at an elevation of more than thirteen thousand feet on Mount Wilson and isolated fifteen miles southwest of Telluride, he confronted San Juan mining at its most difficult.

Out of his experiences came back to back novels in 1902 and 1903—*To the End of the Trail* and *The Blue Goose*, respectively, both set in the Telluride area and involving mining and mining people in their plots. Of the former, the *Telluride Daily Journal* (March 25, 1903) declared that "most of the characters are recognizable to people at all familiar with events in this section of the country the past year or two, and the book will be entertaining and instructive." The same might have been written about the second novel.

Miner Alfred King knew the dangers of the industry better than anyone. While working in the Calliope Mine in March 1900, three boxes of giant caps exploded near him, blinding him. Following his recovery, he toured on the Chautauqua circuit—lecturing, reciting his poems, and playing the "flute beautifully." A popular figure in the Colorado mining regions, King came to be called the "Milton of Colorado—the blind poet of Ouray."

King captured the miner's work and dangers in his poem, "The Miner," which concludes:

> *Thus the battle he fights for his daily bread;*
> *Thus our gold and our silver, our iron and lead,*
> *Cost us lives, as true as our blood is red,*
> *And probably always will.*

He published two books of poetry after his accident and recovery, *Mountain Idylls* (1901) and *The Passing of the Storm* (1907). The first contains short poems, the latter an epic-length one of the same name. Said a *Denver Times* reviewer of King's first book, applauding the expression of what he felt, "The verses have much merit, and indicate the soul poetic of the writer."

A third literary figure tried to capture the San Juans of an earlier time. Harriet L. Wason arrived in Silverton in 1875. She spent most of her life on a ranch near the future Creede, and the rush to that district happened in her backyard, with part of Creede carved out of her land. The best of her poetry deals with the San Juans and Colorado of her era.

Her first volume of poems, *Letters from Colorado*, appeared in 1887. It was followed by another book and eventually other poems. She was hailed as a "favorite Colorado" writer. One reviewer glowingly described the experience of reading her poem "The River of Souls": it "is to have one's spiritual strength renewed. The romantic element in her poetry gives her a strong hold on some readers." In other poems, many of which do not deal with mining, she wrote about the recently discovered cliff dwellings on the Mancos River.

Few people remember these three authors today. Nason and King wrote nothing else about their San Juan experiences, while Wason continued writing until she died in 1904. Each presented a view of the San Juans they knew and caught the spirit of the time and place.

Perhaps Creede newspaperman Cy Warman may have captured the spirit of the age best of all in his short poem about his town at the peak of its boom days. It contains what may be the most famous lines ever written about a San Juan community.

> *Here's a land where all are equal—*
> *Of high or lowly birth—*
> *A land where men make millions,*
> *On mineral mountains feed.*
> *It's day all day, in the daytime,*
> *And there is no night in Creede.*

An unnamed Silverton poet opened the New Year in 1903 with a six-stanza poem for readers of the *Standard*. In this individual's eyes, local history had started to assume historic proportions. After all, Charles Baker and the men with him, whom the author refers to, had arrived way back in 1860.

> *When Baker, from yon mountain top,*
> *Beheld the spot that bears his name*
> *He did not dream that here one day*
> *A splendid city would acclaim*
> *The riches at his feet.*
> *. . .*
> *And when the throng of eager men*
> *Men of heroic mould and true*
> *Wrought mines that silver might be had*
> *They builded better than they knew*
> *These men now gone.*

By now, as war started to engulf the world, San Juaners looked back from 1914 to the generation of pioneers who had opened and settled those beautiful mountains and valleys they now called home. They might have remembered more than actually happened in what was becoming in their eyes a heroic epic.

Legend and history, fact and fiction, were becoming intertwined in their stories and their minds. Newspapers were printing stories about olden times, say forty or fifty years ago. Old-timers liked to recount stories about a romantic time before the modern age set in. Some people had started thinking about writing histories of those frontier days of yesteryear. To return to our old friend Mark Twain, writing now in his later, more cynical years, "A historian who would convey the truth has got to lie. Often he must enlarge the truth by diameters, otherwise his reader would not be able to see it."

The captain of Telluride's "champion baseballists says his club is just dying to do up either Rico or Ouray" (Evening News, *July 21, 1884*).

CHAPTER TWELVE

"Take Me Out to the Ball Game"

*Whoever wants to know the heart and mind of America had
better learn baseball, the rules and realities of the game.*
—Jacques Barzun, *Baseball Quotations*

*I don't have to tell you that the one constant through all the years
has been baseball. America has been erased like a blackboard, only
to be rebuilt and then erased again. But baseball had marked
time with America . . . It is a living part of history, . . .
It continually reminds us of what once was . . .*
—W. P. Kinsella, *Shoeless Joe* (1982)

Baseball, as both of these writers remind us, was, and is, a vigorous player in the American story and a living component in the American character. Certainly there existed no question about that a century or so ago. In the San Juans, in the late nineteenth and early twentieth centuries, baseball held center stage.

Town teams, not professional teams, were the fans' favorites, favorites they supported with their cheers, their turnout at home and away games, and their pocketbooks that fattened or withered with the outcome of every game. Betting on the local nine was a primary pastime of the game.

While the game itself would be familiar to today's fans, some of the terminology would not. The loyal supporters would have been called cranks or rooters, the pitcher, a hurler, and the outfielders left-, mid-, or right-scouts. A run was a tally or an ace, the infielders, base-tenders and the batter a striker. The shortstop would have been a short scout and the manager a captain who usually played.

San Juaners could watch or participate in and bet on other sports—foot racing, firemen's races, drilling contests, and eventually basketball and football—but none ever became as popular or universal. The local nine represented the community as nothing else did; gloom, despair, and agony reigned at the game's end, much as it did in depressed Mudville when "mighty Casey struck out."

Urbanization had only barely taken hold in the San Juans when baseball arrived. Allen Nossaman, in the first volume of his *Many More Mountains*, recounts the story of the July 4, 1876, Silverton celebration, which included horse and foot races and a baseball game. Sadly, "no list of participants has ever been found, nor is the outcome known."

A game on the Fourth was almost a given, and in this day and time before town teams appeared, it was probably a pickup game between local players rather than regular teams. Certainly an 1879 Silverton game proved to be that sort when it ended with a final score of 36–24, a high-scoring result not all that unusual. With players wearing no gloves playing on a field that it was hoped might be somewhat flat, but not manicured, nor even cleared of all the rocks, such a result had to be expected.

Meanwhile, Lake City's *Silver World*, without elaborating, noted on October 16, 1875, that "baseball mania has a few victims here." The next year the paper enthusiastically announced that a baseball club had been organized, along with practice games, and the development of a "sufficient number of good players for a first class nine." Having reached this point, the editor, in May, went to a practice game and received a "pleasant surprise." A "complete set of caps (white corded with blue)" had arrived for the first and second nines.

This too was typical. The first nine would represent a community against other teams when challenged or challenging. Not only did the town's self-image benefit, but according to the same paper, on April 18, 1876, the players did also by taking to the baseball diamond for exercise to benefit their health. "The spring season opened and the boys feel the need for more exercise in order that the digestive organs may not be impaired." The medical veracity of this may be doubted, but not the fact that the game was played in the "neighborhood of the beer garden where refreshments were handy." Whether that improved the digestive organs has been lost to history.

During the 1880s, the San Juans matured, and baseball spread throughout the region. Games were played in large towns and small camps, not only on holidays but also on weekends, as travel became easier, thanks to the railroad and better roads replacing pioneering trails. Baseball teams took to the field at elevations that would shock purists of the game who fret even about records the Colorado Rockies set at mile-high Coors Field today.

"Sunday baseball is preferable to the average Sunday sermon," intoned crusty Dave Day. He also thought in spring a young man's fancy turned to baseball.

When the Red Mountain nine played to a 24–24 tie, called at the end of six innings, on neighboring Ironton's field, the elevation topped 9,800 feet. In spite of that, it was a great game; apparently no one complained about balls soaring too far in the light air. Telluride at 8,500 feet, Ouray at 7,800 feet, and Silverton at 9,302 feet were three San Juan baseball hotbeds, and nobody whined or even seemed to pay attention to the altitude. The Tomboy team later practiced at more than 11,000 feet and then went down to the lower elevations to play Telluride.

Silverton might have been a baseball hotbed, but it had trouble getting started in the spring of 1888. The *San Juan Democrat* (May 31, 1888) chastised locals, noting that there "are plenty of good baseball players in Silverton if the citizens would only give them proper encouragement." The

> *It was said that "no one threw harder than Smoky Joe Wood." He played for Ouray before joining the Boston Red Sox, where he won three games in the 1912 World Series.*

editor's admonition must have worked for within a week, the "club is out practicing" and for the short "time they have been organized are doing very well." So excited were they that the nine challenged Ouray, Telluride, and Durango to games but had heard nothing.

In this day and time before organized leagues, challenges were sent out, usually with a boast about how definitely good the challengers were. Back in July 1884, the following comment appeared in Telluride's *Evening News*: "Capt. Crawford of 'Our Boys' the champion baseballists of the San Juan say his club is just dying to do up either Rico or Ouray."

This was not amateur baseball in the sense that unpaid players took to the field for the love of the game. A little money helped secure a good man or two, and to the victors went the spoils. Silverton, for its July 4, 1881, celebration, offered twenty-five dollars to the winner and the same amount for the foot race, both of which paled beside the two-hundred-dollar purse for the horse race. Durango carried off the money when it defeated Silverton 11–7.

Gambling was part of the excitement in those holiday games and others as well. For Rico's July Fourth game in 1880, there was a "lively interest manifested in selling pools." Those who supported Kelly's nine, a 15–10 victor, took home the money. Win or lose, a "jolly time was had by the assembled crowd."

The 1890s saw an explosion of interest in a variety of sports as well as attention to the importance of exercise. As that popular 1891 volume, *Golden Manual or The Royal Road To Success*, told its readers in no uncertain terms, "Nothing need be said concerning the value of sound health. It is the condition on which all success in life depends." To be sure its readers understood, this bit of advice followed. "A sound mind in a sound body is the first requisite for making the most of yourself and your pursuit."

Physical fitness became the fad of the hour. Americans throughout the country enjoyed both participation and spectatorship. The San Juans, now a quarter of a century old, marched right in step with their lower-altitude cousins.

Nothing was more popular than the bicycle, which crossed into areas of both sport and recreation. Bicycle fever struck America and the San Juans in the early days of the 1890s. One reason for this was the appearance of the safety bike, with two wheels the same size and foot brakes, instead of the

No. 13. THE ONWARD.   **Our Drive Price $42.25**
Order by Number.   Retails everywhere at $75.00 and upwards.

Sent C. O. D. to any one on receipt of $1.00 as a guarantee of good faith. We furnish the same Wheel with best quality 1-inch Cushion Tyre at $39.20.

DESCRIPTION. FRAME, improved 1894 pattern diamond frame, made of steel tubing, with important parts of steel drop forgings; **WHEELS**, 26 inches, corrugated rims with tangent spokes and brass nipples, fitted with 1½-inch Morgan & Wright pneumatic tyres; **STEERING FORK**, made of steel tubing with adjustable nickeled coasters; **HANDLE BAR**, made of steel tubing, fitted with cork grips with nickeled ferrules, properly curved downward and backward, and brought well into position for rider; **BEARINGS**, all high grade steel, carefully hardened, dust proof, full balls to wheels, crank axle, steering head and pedals; **CRANKS**, detachable, 5½-inch throw; **PEDALS**, made of steel, hardened and fitted with moulded rubbers; **CHAIN**, Humber pattern, ¾-inch block chain, 1-inch pitch, true to gauge, **rear adjustment; BRAKE**, improved Plunger pattern; **GEAR**, Sprocket wheels, 17x9 geared to 49 inches; **SADDLE**, improved 1894 Garford style, as shown in cut, with tool bag, inflater, oiler and necessary tools; **WEIGHT**, all on, 36 lbs.; **FINISH**, all bright parts finely finished, japanned with our special enamel which produces the best finish on the wheel which can be obtained. **ILLUSTRATION** shows the bare wheel, but our price includes and we furnish it complete with mud guard, brake, and all attachments. This Wheel is designed for boys from 12 to 18 years of age.

Our Drive Price $42.95.   C. O. D. TO ANY ONE ON RECEIPT OF $1.00 AS A GUARANTEE OF GOOD FAITH.

DESCRIPTION. FRAME, Improved 1894 pattern **for girls**, with detachable bar for boys use, made of steel tubing, with principal parts of steel drop forgings; **WHEELS**, 26 inches, corrugated rims with tangent spokes and brass nipples, fitted with 1½-inch Morgan & Wright pneumatic tyres; **STEERING FORK**, made of steel tubing, with adjustable nickeled coasters; **HANDLE BAR**, made of steel tubing, fitted with cork grips with nickeled ferrules, properly curved downward and backward, and brought well into position for rider; **BEARINGS**, made of high grade steel, carefully hardened and dust proof, full balls to wheels, crank axle, steering head and pedals; **CRANKS**, detachable, 5½ inch throw; **PEDALS**, made of steel, fitted with moulded rubbers; **CHAIN**, Humber pattern, ¾-inch block chain, 1-inch pitch, true to guage, rear adjustment; **BRAKE**, improved Plunger pattern; **GEAR**, Sprocket wheels; 17x9, geared to 49 inches; **SADDLE**, improved 1894 Garford style, as shown in cut, with tool bag, inflater, oiler and necessary tools; **WEIGHT**, All on, 37 lbs.; **FINISH**, All bright parts finely nickeled, japanned with our special enamel, which produces the best finish on the wheel which can be obtained.

We furnish same Wheel with 1-inch Cushion Tyre at $39.20.

No. 14. THE FORWARD.

Order by Number.

The bicycle salesman was a sure sign of spring. Not all were as pricey as these models. Sears Roebuck's prices ranged from $24.95 to $56.50.

ordinary ones of the pre-1890s. With its large front wheel (six feet), a tiny back wheel, and weak brakes at best (sometimes none at all), the adventuresome 1880s rider depended on a long step down or finding a supportive branch or friend to help him safely dismount. Regardless, when snow arrived, bikes hibernated.

Come spring, the bicycles reappeared on the streets. As the *Ouray Herald* announced in January 1896, "The first bicycle drummer to strike Ouray for spring business came in this week. Soon they will be thicker than cigar drummers." In the excitement to ride, some bicyclists left the road and took to the sidewalks. That was a no-no, declared the Silverton City council, which promptly instructed its marshal to keep them off.

Bikes were not inexpensive. The Monarch and Defiance brands cost forty, fifty, and sixty dollars, and the top of the line, the Monarch chainless, claimed a hundred; "perfection is the result of our long experience." If one did not want the bicycle, he could write the company and for twenty cents receive a deck of Monarch playing cards featuring some of the stars of bicycle racing and that darling of the stage, Lillian Russell.

For those disinclined to pay that much for a bicycle, the *Sears, Roebuck* 1897 catalog offered men's, women's, and juveniles' bikes priced from $24.95, not to mention lamps, bells, cyclometers, repair kits, pants, hats, and everything else the bicyclist might need. Baseball bats and balls started at 20¢, mitts at 39¢, shoes at $2.95, and hats at 32¢. Boxing gloves opened at $1.75 and wooden fishing rods at 9¢. The San Juans had plenty of Izaak Walton devotees who nearly managed to fish out some streams.

The San Juans never had professional bike races like those held back east, but several towns organized clubs in the League of American Wheelmen. Better country roads to wheel over were one thing bicyclists desired and worked to achieve. Then there were always those firsts—the first to ride from Ouray to Silverton and so forth. One San Juaner took this to the extreme when he rode from Rico to Silverton on his way to Lake City and then home to Kentucky.

Boxing was becoming popular as well, although many people protested what they saw as a brutal sport. Locals fought for town honor just as the baseball teams did. One match in Creede brought two western legends together. Bat Masterson refereed, and Bob Ford (the "dirty little coward who shot Mr. Howard [Jesse James]") served as timekeeper. Neither did much work; the main event ended with a knockout in the first round.

Wrestling also gained a few devotees, and even handball appeared at Lake City. Drilling contests gained steady popularity. Here contestants hammered with a hand sledge and steel bit, usually into granite rock, in a race against time to see who, or which team, could drill the most inches. Single-jack and double-jack contests drew miners from throughout the San Juans, and often beyond, lured by substantial prizes. Ouray offered $600 for first prize and $200 for second in the 1899 Labor Day double-jack drilling contest. Single jackers received $115 and $75, and there was another contest for boys younger than sixteen. According to the announcement, "it is open to the world," but it was more practical for local miners. Some participants perhaps came from other Colorado mining districts or Utah.

The interesting thing about drilling contests was that the skill was becoming obsolete in many mines by this time. Machine drills had replaced hand drilling in most larger operations.

Meanwhile, back on the diamond, baseball continued on its merry way, though not without some protests. In order to have a better chance to win, some teams would bring in ringers and pay professionals, or semiprofessionals, to play on the local nine. Quite often, this meant a battery of pitcher and catcher. This caused a kick, for example, when Del Norte protested and refused to play because the Creede team had fielded seven players from Denver and two hometown boys. Substitutes were found, but Creede smashed Del Norte 33–9 in a game called after six innings.

Umpiring also triggered its share of grief. Usually the home team furnished the umpire, which might mean trouble in the waiting or, as one newspaper noted, the visitors would "feel they had fallen among thieves." The crisis was sometimes resolved by each team furnishing an umpire. One problem with umpire's decisions came with the betting on the game. When Ironton played Ouray in 1892, the odds stood 5–1 against them. One or two close calls could tip the balance, and shocked bettors would have to fess up their bets.

While Sunday games had never been a problem in the mining districts as they were back in the more puritanical East, a few folks always felt that the Lord's Day should not be desecrated in such a manner. Durango's *Great Southwest* (August 5, 1899) thought that more "proper respect to the religious opinions" ought to be valued. The editor wished "these ball games would be arranged some other day besides Sunday." He was wasting his editorial ink.

Baseball remained popular into the twentieth century. When Ames played Ophir in 1907, the Ophirites won a "very exciting game," 27–18.

A vintage baseball team takes the field where Smoky Joe played. Photo by Jeff Reznikoff, Ouray Times.

Despite this, scores were lower because of better equipment and playing fields. Creede, for example, defeated Monte Vista in a double header, 6–5 and 3–2. Now fans could also follow their favorite National or American League team, at least the scores, in the local newspapers.

Baseball still meant spring. Telluride's *San Miguel Examiner* treated its readers to this little opening season poem on May 18, 1907.

> *In the spring the young girl's fancy*
> *Lightly turns to thoughts of hat,*
> ...
> *In the spring the young man's fancy*
> *Lightly turns to thoughts of ball.*

Very few San Juaners ever made the major leagues, but one did, Ouray's Joe Wood. At the age of fifteen, he played on the town team in 1905. "As a boy I had nothing on my mind but baseball." He not only loved the game; he played it with skill and professionalism. In 1912, Smoky Joe Wood went 34–5 for the Boston Red Sox and pitched his team to a World Series victory with three wins.

High schools now had athletic teams in football, baseball, basketball, and track, all for the males. Women sat on the outside watching. Then, six young ladies from Durango had the courage in 1905–6 to take the issue of women's sports all the way to the school board. They were rewarded by being permitted to have a basketball team, but none of the boys could watch the games for fear of raising those base passions Victorians so abhorred. Ouray also fielded a team. Slightly different from the men's version, women's basketball teams had six players—three offense, three defense—and neither group could cross midcourt.

As the years wound down toward the Great War, San Juan sportsmen and their teams continued playing. Mining might have declined and camps that once fielded a nine could not find enough players anymore, yet interest stayed high. There were now local leagues with teams playing a regular schedule. Companies such as Telluride's Liberty Bell mine and Durango's American Smelting & Refining sponsored some of the nines.

Some things never change. Creede went to play Monte Vista and lost 6–5. The players came home with a "grievance" against the umpire, who "rankly favored his home team." Said the *Candle*, "Umpires who allow prejudice to supplant honor and honesty eventually will kill baseball if tolerated."

San Juaners never lost their interest in baseball nor their desire to win. Perhaps no one has captured the spirit of nineteenth- and early twentieth-century baseball better than Ernest Thayer in his immortal baseball poem, "Casey at the Bat."

> *Oh, somewhere in this favored land the sun is shining bright;*
> *The band is playing somewhere, and somewhere hearts are light.*
> *And somewhere men are laughing, and somewhere children shout:*
> *But there is no joy in Mudville—mighty Casey has struck out.*

It could have been Lake City, Silverton, Rico, Creede, or any San Juan nine in the era when baseball was king.

Indeed, "remember" these people as you pass by.

EPILOGUE

"Remember me as you pass by"

A tombstone epitaph in a mining-camp cemetery contained an oft-repeated first line and poem:

Remember me as you pass by,
As you are now so once was I.
As I am now so you will be
Prepare yourself to follow me.

That might just as easily serve for the San Juaners in this story and for their era and communities as well.

Most of the camps have slipped quietly into becoming ghost towns with a few tumbledown buildings, if any, to mark their passing. The towns linger as mere shadows of yesterday's high mining days with the exception of Telluride, which stopped cussing snow and now lives on as ski Mecca and festival gathering place, and Ouray, which parlayed a beautiful location and hot springs into becoming a tourist haven. Meanwhile, the smelter town and supply and railroad center, Durango, has grown and flourished into the twenty-first century.

It is easy to see that mining once thrived in the San Juans. The mountains bear the scars of the miners having prospected and dug into them in their determined, ongoing search for nature's mineral bounty. From valley floor to mountaintop, their hopes and expectations can be seen in mine portals and mine dumps. Roads and trails going nowhere except to a collapsed mine, or a pile of rocks, tell a tale of expectations floundering against harsh reality.

Victorian buildings and homes tarry in the towns, and along with a scattering of windblown relics in the camps mark the passage of a generation or

Death was a constant companion to young and old.

two of the folks who became Coloradans for a while at least. They remind visitors of what once was and a host of might-have-beens and used-to-bes. A few engines and railroad cars are sitting in parks, and one operating narrow-gauge train from Durango to Silverton recalls the railroad that was so crucial to an era's life and economy. It still booms Silverton during the summer season as it did when it first arrived back in 1882.

Visitors come and look about today. They try to imagine what life once was like when mining news and community gossip filled the day; when ore wagons, mules, and carriages crowded the streets; when legends were made and legends died.

That might be fascinating, but it was the people in these towns and camps who gave heart and soul to this story. Whether their quest proved successful matters little today. They lived, loved, dreamed, worked, prospered, schemed, and failed, with a few succeeding gloriously, as mining ebbed and flowed in the San Juan Mountains of Colorado.

They are gone now, the people who made this story come alive. They have finally put down roots in the cemeteries that are scattered about the mountains and valleys. The gold and silver they sought might never have been theirs, but they left behind a western saga that was part of an eternal legend of man's search for wealth against odds that ranged even higher than those they faced in the gambling hells so many frequented.

Let us now return once more to Colorado poet Thomas Hornsby Ferril, who illuminated this epoch and its people in his poetry as few others have done in their writing. He wrote, in the opening and closing stanzas of his poem, "Judging from the Tracks":

> *Man and his watchful spirit lately walked*
> *This misty road . . . at least the man is sure,*
> *Because he made his tracks so visible,*
> *As if he must have felt they would endure.*
>
> *And judging from the tracks, it's doubtful if*
> *A guardian angel moved above his head,*
> *For even thru the mist it can be seen*
> *That he was leading and not being led.*

INDEX

aeroplane, 41
Ancient Order of United Workmen, 113
Animas Forks. See individual subjects
automobile, 41, 54

bachelor. See individual subjects
Backus, Harriet, 57, 58–59, 63, 107, 139
bands, 140
banks, 5
baseball, 145–48, 151, 152–53; umpires, 152, 154
Beaumont Hotel, 53
bicycles, 148, 150, 151
Botkin, Alex, 140
boxing, 151–52
breweries, 6–7
Brinker, John, 105
Bronco Lou, 124
Brunot Agreement, 44
Bryan, William Jennings, 25
burros, 22, 43, 76
businesses, 3, 5
businessmen, 1–2, 6, 64–67; city government, 14–18, 22
businesswomen, 1–2, 9
Byron, Lord, 57

Camp, Alfred, 42–43
Capitol City. See individual subjects
cats, 76, 98
Chautauqua, 135
children, 74, 76–79; athletics, 73; red-light district, 132; school, 69–70, 73
Christmas, 77
circus, 76–77
Cornish choir, 140
cost of living, 63–65, 67
Crash of 1893, 11
Creede. See individual subjects
crime, 19

dance halls, 126, 128
Darley, Alex, 97, 104
Darley, George, 67, 92, 121; minister, 94–96, 97, 98, 101
Day, David, 51, 67, 105, 124, 125; baseball, 147; career, 28–31, 35; circus, 76–77; railroads, 52
death, 92
Democrats, 25
dentists, 86, 88, 90
Denver, 36
Denver & Rio Grande Railroad, 47, 49–52, 53–54

Dickinson, W. B., 43
diseases, 90–91
doctors, 81–82, 86, 88
dogs, 17, 32, 76, 98
drilling contests, 152
drugs, 129–30
Durango. See individual subjects

elections, 22
electricity, 8
environment, 19
Eureka. See individual subjects

Ferril, Thomas Hornsby, xi, xiii, 90, 159
fire, 19–22
Ford, Bob, 151–52
Fort Lewis, 52
fraternal lodges, 110–13

gambling, 129, 148
Gibbons, James, 96, 100–101, 102, 116
Gibbs, Martha, 69
Good Templars, 118
Government (town), 13–14, 21, 25
Grand Army of the Republic, 112, 113, 115–16
Greeley, Horace, xi

Hartman, Magg, 96
high school, 154
Hoffmann, Ernie, 128, 129
hospitals, 87–88
hotels, 8, 9
hot springs, 88
Hough, Frank, 90
Howardsville. See individual subjects

illnesses, 81, 86–87
Improved Order of Red Men, 115
Ironton. See individual subjects

Jackson, William Henry, 42

Keeley Institute, 129–30
Kellogg, Walter, 74
King, Alfred, 142
Kingsley, Charles, ix
Knights of Pythias, 115

Lake City. See individual subjects
League of American Wheelmen, 151
library, 17, 136, 137
literary societies, 116
Livermore, Robert, 126, 129
Long, Ruth, 140

mail, 45
marshal, 17–18, 25
Masterson, Bat, 151
Mears, Otto, 44–45
medicine, 82–84, 92
Mencken, Henry L., 27, 34
Methodists, 107
Mineral Point. See individual subjects
Miners' Library Association, 116
mining, xi–xii
minstrel company, 140
Montgomery Ward, 63
Mott, Mary, 74
movies, 7–8, 139–40
mules, 43

Nason, Frank, 141–42, 143
newspapers, 27–32, 34–35, 38, 74
Nossaman, Allen, 146

Olcott, Eben, 9, 77, 107
opera house, 135, 136
Ophir. See individual subjects
opium, 32, 86
outhouse, 32

Paddock, Annie Laurie, 69, 71–72
Palmer, William Jackson, 47
Parrott City. See individual subjects
patent medicines, 83–86
Pinkham, Lydia, 86
Populists, 25
Prohibition, 105
Propper, Gideon "Gid," 31
prostitution, 96, 124–26. See also red-light district
Pure Food and Drug Act, 130

railroads, 41, 47, 48–53, 158
red-light district, 6, 120–24, 128–29, 130–31, 132
Red Mountain. See individual subjects
Republicans, 22
Rico. See individual subjects
Rio Grande Southern Railroad, 52, 54
roads, 43–45, 54
Rockwood. See individual subjects
Rockwood, Thomas, 29
Rocky Mountain Tourist, 62
Romney, Caroline, 31–32, 35
Roosevelt, Theodore, 25
Rose's Cabin. See individual subjects

salesmen, traveling, 9–10
saloons, 124, 126, 131
schools, 70, 71–72
Sears, Roebuck and Company, 63, 64, 77
Sherman. See individual subjects
Silverton. See individual subjects
Silverton, Gladstone and Northerly Railroad, 52
Silverton Northern Railroad, 52
Silverton Railroad, 52
snowshoes (skis), 43
stagecoach, 45–46
Sunday, 97–98
Sunday, Billy, 95

Tabor, Horace, 46, 136
teachers, 70–72, 74
telephone, 8
Telluride. See individual subjects
Thayer, Ernest, 154
theaters, 129, 137, 138
tourism, 53–54
Turner, John, 29
Twain, Mark (Samuel Clemens), 6, 28, 79, 88; on history, 143; opera, 135; saloon keeper, 126; whiskey, 132

urbanization, 65
Utes, 44

vaudeville, 139

Warman, Cy, 142–43
Wason, Harriet L., 142
Watkins, Waltus, 88
Webster, Daniel, 13
Western Federation of Miners, 37, 119
White, William Allen, 27
Williams Tourist Guide, 62
Woman's Christian Temperance Union, 79, 86
women, 8, 90, 122; role of, 58–62, 102, 104; teachers, 72
Wood, Smoky Joe, 148, 149, 153
World War I, 11
wrestling, 152

www.ingramcontent.com/pod-product-compliance
Lightning Source LLC
Chambersburg PA
CBHW082121230426
43671CB00015B/2771